Methods In Climatology

Victor Conrad

METHODS IN CLIMATOLOGY

LONDON: GEOFFREY CUMBERLEGE

OXFORD UNIVERSITY PRESS

METHODS IN
CLIMATOLOGY

BY

VICTOR CONRAD
HARVARD UNIVERSITY

CAMBRIDGE, MASSACHUSETTS

HARVARD UNIVERSITY PRESS

1946

Printed in the United States of America by
Lancaster Press, Inc., Lancaster, Penna.

TO

IDA CONRAD

MY COMRADE IN LIFE AND WORK

THIS BOOK IS DEDICATED

PREFACE

CLIMATE influences the surface of the earth, and this conversely, in its varieties, determines the climate under otherwise equal conditions. This intimate mutual connection makes climatology and climatography appear as parts of geography, because they are essentially necessary to describe the surface of the earth and its changes. These ideas find their expression in the fact that generally in colleges and universities, climatology as a whole is treated in the geographical departments. Perhaps the dependent role of climatology may be attributable also to the fact that geographers have so greatly furthered this science.

For the most part, climatographies have been written by geographers. Therefore, geographical methods are kept in the forefront, and specifically climatological methods are not so much used. It would be satisfying if this book offered a bridge connecting the two realms.

To judge from the author's many years of experience in Europe, followed by a few years in the United States, the student in geography has perhaps sufficient knowledge of climatology, but he does not know how to deal with the original data of observations. He is, I should say, familiar with the results but not with the ways of getting them. This holds, particularly, for the different methods of mathematical statistics which can be adapted to an individual representation of the climates; only thus can too-schematical descriptions be avoided.

This book should make the student acquainted with a number of methods of mathematical statistics and theory of probability applicable at once to climatological problems. Approximations are used as much as possible in order to facilitate arithmetic. The student who is interested in the mathematical proofs and derivations as well as in their philosophical background must be referred to the special technical literature.

The general introduction presents climatology as a world science, and its international organization. The number of observations in the meteorological register makes the necessity of statistical methods evident. Their special application to climatological series is discussed. An instruction, easy to understand, is given for computing periodical phenomena by means of harmonic analysis.

In the foregoing, the overwhelming role of the frequency distribution has been emphasized. This idea led L. W. Pollak to introduce a new system of treating climatological observations by means of automatic statistical machines. This system covers the entire body of the observational data and may represent the future of rational climatological statistics. Pollak's method is not discussed here, however, because only with the resources of great institutions is it practicable.

The general representation of climatological elements, factors, etc., is followed by discussions of the special characteristics of the single elements.

The radiation elements are not dealt with. As far as frequencies and different correlations with other elements are concerned, the methods are identical with those applied to other climatological elements, on the one hand. On the other, such a discussion would have had to include the physical and astronomical nature of the subject. This exceeds the range of this book.

The first two parts of the present book are concerned with the variations of the elements in the course of time at one fixed place. The third part presents the comparison of the elements which are observed synchronously at different places, and arrives at their geographical distribution.

Critical scrutiny of the observational data leads to the examination of the homogeneity of climatological series and to the reduction to an identical period   These problems remind me of Hann's quotation of Francis Bacon, the great English philosopher: *Si quis hujusmodi rebus ut nimium exilibus et minutis vacare nolit, imperium in naturam nec obtinere nec regere poterit.*[1]

These efforts are of fundamental importance not only for the climatologist but also for everybody who is interested in climatological series, as are the long-range forecaster and those dealing with periodicities and tree-growth analysis.

Comparisons of data at different places reveal the concept of coherent and non-coherent climatic regions and of the climatic divide. These ideas show also clearly that the theory of correlation in the realm of climatology should play a significant role. An intermediate chapter deals with this subject.

[1] "A scholar who is unwilling to take pains over such investigations because they seem too insignificant and microscopic will be able neither to gain nor to maintain mastery over nature "

Climatological examples demonstrate the computations so that everyone with high-school training should be able to understand the mathematical procedure. Linear correlation, simplifications of the calculation, and regression equations are discussed. Finally, the formulas of partial correlation of three variants are added. Rather much space is devoted to graphical methods of representation, especially in connection with mountainous regions. The great importance of anomalies and isanomals is shown and supported by examples. At the end of this chapter, air-mass climatology, continentality, etc., are mentioned.

The fourth section gives suggestions for the arrangement of a more or less complete climatography.

In the appendix, models for climatic tables are presented, a table for easy calculation of the probable error, an auxiliary table for computing the equivalent temperature, and a table giving the numbers of the consecutive days of the year.

The content and order of treatment in this book reflect, for the greater part, the experience the author acquired in his lectures in climatology and in supervising the theses of his pupils at the University of Vienna, Austria. A number of the examples are taken from the author's papers

It is with the greatest pleasure that the author expresses his gratitude to Professor Charles F. Brooks of Harvard University for his repeated encouragement to summarize in book form the methods of climatology. Thanks are also due to Dr. Brooks for his reading of the manuscript and many suggestions as well as for constant help in facilitating the routine work connected with this book.

The author wishes to express his appreciation for financial assistance from gifts of the Associates of Physical Science and Anonymous Fund 15 of Harvard University.

He is deeply indebted to Professor L. W. Pollak, Dublin, Ireland, who himself has contributed so much to the problem of statistical methods in meteorology and climatology, for his reading the manuscript and his valuable comments and suggestions.

His warmest thanks are extended to Professor W. M. Fuchs, New York, Dr. J. T. Morley, Dublin, Ireland, and Dr. C. Chapman, Boston, who did their best to smooth his English. Gratefully he acknowledges the assistance of Miss Helen Gilman for redrawing the figures. Mr. R. W. Burhoe, Acting Librarian of

Blue Hill Meteorological Observatory, provided the necessary literature, always very helpfully and promptly.

Owing to the present war, it has been impossible to obtain permission from the respective scientists and publishers to reproduce here a number of graphs. The author offers his sincere apologies for this omission and assures all concerned that, in normal times, he would have waited until he had received permission, before publishing. In this book, he has at least been careful to see that in the legend every figure taken from another work has been duly attributed to the scholar who originated it.

V. C.

Cambridge, Mass.
October 1, 1944

CONTENTS

PART II

REPRESENTATION OF CHARACTERISTIC FEATURES OF DIFFERENT ELEMENTS

PART III

METHODS OF SPATIAL COMPARISON

TABLES

ILLUSTRATIONS

METHODS IN CLIMATOLOGY

INTRODUCTION

CLIMATOGRAPHY describes the climate, climatology explains it. Climatography prepares the raw material supplied by observation; it is the foundation of climatology. This seems a vicious circle, because it is hardly possible to make a climatography without a knowledge of climatology. This is the present state in which the two branches of science supplement each other. In its early beginning, climatography was an occasional description in words and pictures by travelers, chroniclers, writers, poets, and painters.

The invention of the thermometer and its regular use for observing air temperature was the first step toward a quantitative climatography. Galileo Galilei invented the thermometer in 1597. It became an accurate physical instrument in 1780 when de Luc introduced mercury as a thermometrical substance. Rain gauges are supposed to have been used in India in the fourth century B.C. The oldest regular quantitative measurements of rain were made in Palestine in the first century A.D. With these observations at hand, it has been possible to compare the hydrometeoric climate of ancient Palestine with the present one in an exact way. The invention and development of other instruments has furthered regular quantitative climatological observations. Elements which cannot be observed by instruments are estimated by means of arbitrary scales

A single observation, picked from a series, has usually no climatological meaning. Only the chronological arrangement and, generally, the combination of observations, are sufficient basis for computing average variations and average states. A single observation can be compared to a single exposure of a movie film, which has no real meaning until it is combined in its proper sequence with the rest.

The nature of our subject determines the necessary analytical and synthetical methods. First of all, the meaning of *climate* may be explained:

Climate is the average state of the atmosphere above a certain place or region of the earth's surface, related to a certain epoch and consider-

1

is subject.[1]

Observations are made at isolated points. Only by comparing these data can the climate of the whole region be interpolated. *It is therefore the principal and fundamental aim of climatological methods to make the climatological series comparable.* The more fully this goal is approached the more reliable are the indications of a climatography.

Long-range forecasts or investigations in hidden periodicities depend even more on reliability and exactness, and therefore upon the comparableness of climatic series. Only with comparable climatic series can true averages, true variations, true probabilities be determined and used for making further estimates and inferences.

[1] V. Conrad, "Die klimatischen Elemente und ihre Abhängigkeit von terrestrischen Einflüssen," *Handbuch der Klimatologie*, ed by W. Koppen and R. Geiger, vol I B (Berlin, 1936).

PART I

GENERAL METHODS

CHAPTER I

CLIMATOLOGICAL ELEMENTS. COMPARABLENESS OF CLIMATOLOGICAL SERIES

CLIMATOLOGICAL ELEMENTS result from the analysis of the state of the atmosphere. The combination of all elements occurring at a given moment makes the weather; the average weather means the average state of the atmosphere, that is, the climate.

It is not possible to enumerate all the elements, because their number can be increased arbitrarily, but the following may be listed:

1) Radiation of sun and sky
2) Temperature of the air and of the surface of the earth
3) Wind direction and velocity
4) Humidity and evaporation
5) Cloudiness and sunshine
6) Precipitation
7) Snow cover
8) Air pressure, because of its intimate relation to the instantaneous and average state of weather and atmosphere.

The difficulty is that these items do not represent single elements, but groups of elements. The radiation of sun and sky, as an example, is divided into two main groups: direct radiation from the sun, and radiation from the sky. Each group can be subdivided into any number of elements. There is the total radiation, including the energy of the entire spectrum, which can be divided into the elements characterized by the energy of parts of the spectrum, for example the infrared, the visible, the ultraviolet portions. Other radiation elements deal with the transmission coefficients of certain wave-lengths, with the turbidity factor, the blueness of the sky, photometrical and photochemical brightness, atmospheric back radiation, range of the ultraviolet spectrum, etc.[1]

The group of radiation represents only one example. The same conditions exist with all the other groups. This nearly infinite multifariousness is no misfortune. On the contrary, a clever climatographer, while utilizing the usual elements, should define new elements which give the best, the shortest, and the most accurate contribution to the description of the climate in consideration.

For the present purpose, it is perhaps of greater importance to classify the elements from another point of view.

1) Primitive elements, directly observed or estimated
2) Combined elements
3) Derived elements.

Examples of group 1, *primitive elements*, are temperature, precipitation, wind velocity, wind direction, cloudiness, etc. Examples of group 2, *combined elements*, are turbidity factor, continentality, equivalent temperature, potential waterpower, etc.

There are also elements which may be taken both as primitive and combined elements; examples are relative humidity and cooling power. On the one hand, each can be read directly from its appropriate instrument; on the other hand, each can be calculated by means of a combination of two or three other elements.

The third group is obviously the most extensive and offers much scope to the imagination of the climatographer and climatologist. Only a few examples may illustrate the importance of this group. There are the variabilities of the different elements, the length of certain spells, as of high and low temperatures, of dry or wet weather, etc.; the duration of characteristic average states of the atmosphere—for example, the duration in days of the period during which the average temperature is below the freezing point (frost period) or the duration of the vegetation period, that is the time during which the average temperature is $\geqq 43°F$; the lapse rate of temperature; the manifold use of frequencies and probabilities, anomalies, etc.

From the classification of the elements as primitive, combined, and derived, their infinite multiplicity is obvious;[2] hence the need

---

[1] For the multiplicity of radiation elements, see the monograph of F. W. P. Götz, *Das Strahlungsklima von Arosa* (Berlin, 1926) From the same point of view, W Mörikofer, "Meteorologische Strahlungs Messmethoden," in Emil Abderhalden, *Handbuch der biologischen Arbeitsmethoden* (Berlin, 1920—) Abt II, pt 3, is of interest.

[2] In the foregoing series of climatic elements, the one or the other is enumerated which may be unknown to the reader. Some of them are explained in the following

of some methodical hints on handling these numerous expedients and selecting the best in order to get the most concise and exact description of the climate.

Observations do not always reproduce natural conditions in the right way. There are many sources of error. An appropriate critique of the data should eliminate the systematic as well as random errors. These corrections concern a series of observations made at one place for a period of time, and a comparison of observations made simultaneously at different places.

The preparation of original observations for purposes of comparison, whether as single values or as a series, is one of the main tasks of climatological methods.

I. 1. CLIMATOLOGICAL COLLABORATION AND ORGANIZATION. STATIONS OF DIFFERENT ORDERS. THE "RESEAU MONDIAL"

Climatology is a world-wide science. That is clear from the standpoint of comparability. The observations of a station A have to be compared with those of station B south of A; the observations of C west of A should be compared either with A or with B, and so forth, for all directions of the compass. (See Fig. 1.) Thus we may understand why Sir Napier Shaw writes in his *Manual of Meteorology* (I, 160)."No country, not even the largest, is self-sufficient in the material for study, because that material must be co-extensive with the whole world."

The first series of observations were made by private men and institutions interested in weather and nature. In 1781, the meteorological organization the "Mannheimer Akademie" was founded; and this was the first step toward a world-wide climatography. In their publications, *Ephemerides Societatis Meteorologicae Palatinae*, the results of observations for two American places were also published.

Nowadays, in civilized countries, the State has organized central institutes which administer and extend the network of climatological (meteorological) stations. The expression "network" is taken from the "network of stations" upon which the geodetic survey is based. Figure 1 indicates more fully the meaning of the

chapters in so far as they are of methodological interest Other explanations can be found in V. Conrad, *Fundamentals of Physical Climatology* (Harvard University Press, 1942) and in V Conrad, "Die klimatologischen Elemente und ihre Abhängigkeit von terrestrischen Einflüssen," Köppen-Geiger, *Handbuch der Klimatologie*, vol. IB (Berlin, 1936).

term. The observations of each station have to be comparable with those of the others. The aim of the study of the reciprocal relations between the climatic elements at different places is, on the one hand, to determine the circulation of the atmosphere (physics of the actual state of the atmosphere), and, on the other, to get full information about average and extreme values of climatological elements—at least on the earth's surface (physics of the average state of the atmosphere).

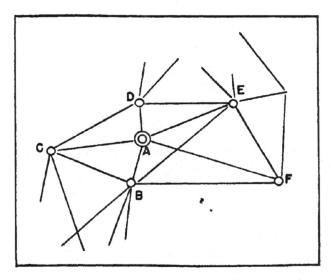

FIG. 1. Structure of a climatological network

For these large-scale purposes, a network, internationally administered, has to cover the entire globe. Regional meteorological efforts found their international organization at a relatively early stage. An international conference of maritime meteorology was held at Brussels in 1853, ninety years ago, and the first international meteorological congress at Vienna (Austria), in 1873, was attended by official delegates from a large number of countries. On this occasion an "International Meteorological Committee" was elected in order to assure international meteorological and climatological collaboration without interruption during the interval between the congresses.

This international organization has the following duties:

1) To make regulations and offer advice concerning comparable observations.

2) To establish definitions of phenomena like rime, snow-cover, day with thunderstorm, day with precipitation, etc.

3) To make regulations regarding the organization of a network for a country. Such a network should consist of stations of different orders:

(*a*) A *Central Office*, or *Central Institute*, is the chief office entrusted by the Government with the management, collection and publication of the meteorological observations of the country [e.g. the U. S. Weather Bureau].

(*b*) A *Central Station* [Section Center] is a subordinate center for the management and collection of observations from a certain province [state; e.g. the U. S. Weather Bureau, Boston, Mass.].

(*c*) A *Station* of the *First Order* is an observatory in which, without the collection of observations from other stations, meteorological observations are conducted on a great scale, i e. either by hourly readings or by the use of self-recording instruments.

(*d*) *Stations* of the *Second Order* are the stations where complete and regular observations on the usual meteorological elements, viz., pressure, temperature and humidity of the air, wind, cloud, rain and hydrometeors, etc , are conducted.

(*e*) *Stations* of the *Third Order*, finally, are the observing stations, where only a greater or less portion of these elements are observed.[3]

A knowledge of the international organization, of its regulations and its definitions, is of great importance to anyone who has to use the meteorological records of other countries. The most essential resolutions and recommendations of the International Meteorological Committee during the years from 1872 to 1905 are summarized in the International Meteorological Codex.[4] In spite of the age of this report (published in English by the British Meteorological Office in 1909), the recommendations are valid and valuable today with the exception of the sections on aerological observations and on weather maps and forecasting. A Spanish version was published by the Observatorio Central de Manila in 1913, and includes the resolutions adopted between 1872 and 1910.[5]

Another merit of the International Meteorological Organization is the publication of the *réseau mondial*, a global network of meteorological stations. This title represents the climatological and meteorological ideal: a network (as stated above), which

[3] Sir Napier Shaw, *Manual of Meteorology*, I (1926), 163
[4] Prepared by G. Hellmann and H H Hildebrandsson.
[5] Sir Napier Shaw, *Manual of Meteorology*, I, 166. For further information, see: Official Reports of the International Conferences.

covers the entire surface of the earth, administered in a uniform manner and devoted to all of mankind. The publication in question is only the beginning of an ideal as yet unaccomplished. It was a great scientific advance when the British Meteorological Office initiated it in 1911. Monthly averages of pressure, temperature, and precipitation at about five hundred places are published for each year. The stations are grouped in 10°-zones of latitude. It is noteworthy that meteorological data can be supplied for all the zones from 80°N to about 61°S. South of this latitude no permanent meteorological station is being operated.

The data and elements given for each station in the *réseau mondial* for the months and the year are:

1) Latitude
2) Longitude
3) Height (meters)
4) Pressure (millibars):
   (a) Mean at the level of station
   (b) Mean at mean sea level (M.S.L)
   (c) Differences from normal
5) Temperature in the absolute scale, starting from 273°C
      below the freezing point of water:
   (a) Mean maximum
   (b) Absolute maximum with date of occurrence
   (c) Mean minimum
   (d) Absolute minimum with date of occurrence
   (e) Arithmetical mean of (a) and (c)
   (f) True average temperature
   (g) Difference from normal
6) Precipitation:
   (a) Monthly amount in millimeters
   (b) Difference from normal

Another document of the international development of climatology and meteorology is the *World Weather Records*, assembled and arranged by H. Helm Clayton, in consequence of a resolution of the International Meteorological Conference adopted at the meeting of Utrecht in 1923. This publication [6] gives for each year of record the monthly and annual values of air pressure, temperature, and precipitation, and averages of all, for about 390 places.

[6] Smithsonian Miscellaneous Collections, vol. 79 and vol. 90, Washington, D C. 1927 and 1934.

The great advantage is that the publication offers long and, as far as possible, homogeneous series, which overlap one another for the period of observation. Thus the work is indispensable if variations of the three elements have to be compared.

I. 2. THE METEOROLOGICAL REGISTER

The meteorological register contains the records of the original, regular observations made at the climatological station. All climatological studies are based ultimately upon the registers of the stations.

The utmost international uniformity of the registers would be advantageous for all climatological investigations which cover a region larger than one national network. This was the reason why the International Meteorological Organization (Utrecht, Holland, in 1874) suggested a special form for the register of stations of the second order. Figure 2 shows this international form with a few supplements. As far as the units of pressure and altitude, the temperature scale, etc., are concerned, uniformity could not be achieved.

This form, which assumes that the elements mentioned are observed three times a day, is used by most of the central institutes of Europe, of the U.S.S.R., etc. In many networks, the register is more or less modified, according to the various conditions and requirements. It is to be hoped that these modifications will be restricted more and more, and will give way to complete uniformity.

As far as the organization of the United States Weather Bureau is concerned, J. B. Kincer may be quoted: [7]

The Weather Bureau has maintained some 200 observing stations known as stations of the first order, manned by professional meteorologists, at which complete meteorological observations are made, including automatic, continuous instrumental records of many weather elements. Since these stations are not spaced closely enough to provide data adequate for climatic purposes, a large number of cooperative stations . . . are maintained to obtain the necessary basic data. [The number of these coöperative stations is about 5,000.] On the other hand, the U. S. Weather Bureau has branches in the States which administer the respective climatic section and are equipped with many instruments, especially automatic-recording instruments.

[7] "Climate and Weather Data for the United States," in *Climate and Man* (Yearbook of Agriculture, United States Department of Agriculture, Washington, D. C., 1941), p. 689.

INTERNATIONAL CLIMATOLOGICAL REGISTER

Name of the place............ ....Country (State)......... ....

lat. = ...long = ...elev.... .... Meridian of time.. ... .

Year........... ... ... Hours of observ ....

Month...... .. ..

(*) h_B = ...

(†) h_T = ...

(‡) h_R ....

Day	Pressure units			Dry bulb (°F) (°C)				Wet bulb (°F) (°C)				Vapor pressure units			Relative humidity per cent		
	h	h	aver.	max	min	h	h	aver.	h	h	h	aver	h	h	aver.	h	h	aver.
1																		
2																		
3																		
4																		
..																		

Day	Cloudiness (0–10)			Wind— dir. and vel (Beaufort)			Precipitation observed at ...h units	Kind of precipitation	Depth of snow on ground units	Remarks about hours and duration of precipitation, thunderstorms, storms, optical phenomena, etc.
	h	h	aver.	h	h	aver vel				
1										
2										
3										
4										
.										

(*) Height of the mercury in the cistern of the barometer above sealevel

(†) Height of the bulb of the thermometer above the ground

(‡) Height of the rim of the receiver of the raingauge above the ground

FIG. 2. Blank form of the international climatological register

Because the network of the cooperative stations of the U. S. Weather Bureau is one of the greatest in the world, it may be convenient to give here the register of these stations as an example of a somewhat modified international form (Fig. 3).

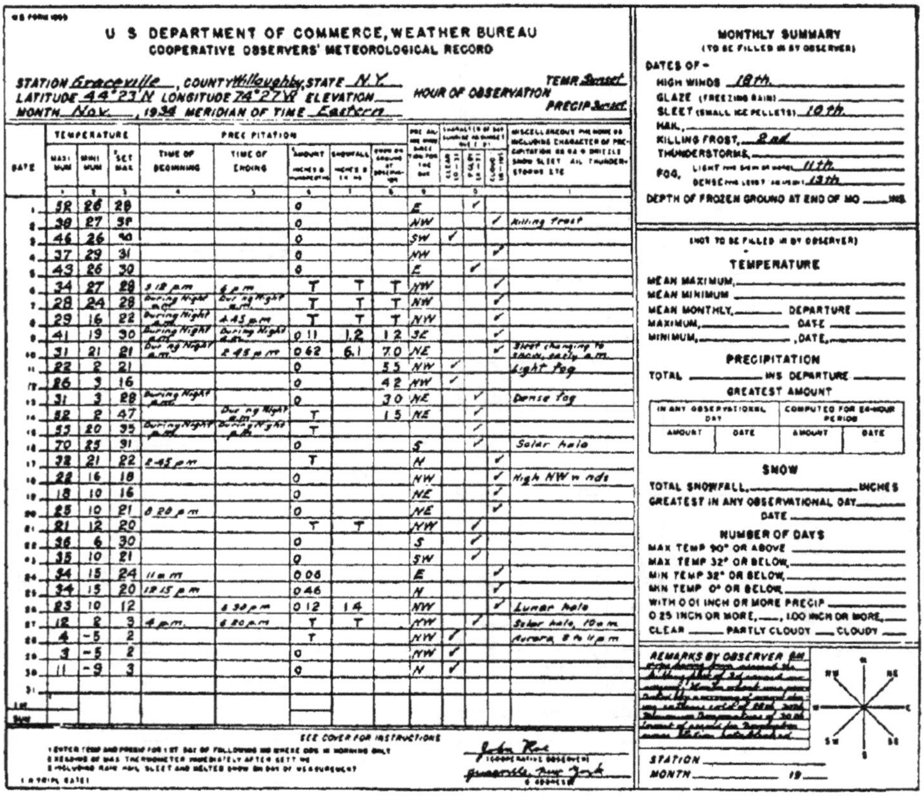

U S DEPARTMENT OF COMMERCE, WEATHER BUREAU
COOPERATIVE OBSERVERS' METEOROLOGICAL RECORD

STATION Graceville, COUNTY Willoughby, STATE N.Y.
LATITUDE 44°23'N LONGITUDE 74°27'W ELEVATION ____ HOUR OF OBSERVATION TEMP. Sunset PRECIP. Sunset
MONTH Nov., 1934 MERIDIAN OF TIME Eastern

DATE	Temperature: Maximum	Minimum	Set Max	Precipitation: Time of beginning	Time of ending	Amount	Snowfall	Snow on ground	Wind direction	Clear	Partly cloudy	Cloudy	Miscellaneous phenomena
1	58	26	28			0			E		✓		
2	38	27	38			0			NW			✓	Killing frost
3	46	26	40			0			SW	✓			
4	37	29	31			0			NW			✓	
5	43	26	30			0			E		✓		
6	34	27	28	3 18 pm	6 pm	T	T	T	NW			✓	
7	28	24	28	During Night a.m.	During Night a.m.	T	T	T	NW			✓	
8	29	16	22	During Night a.m.	4 45 pm	T	T	T	NW			✓	
9	41	19	30	During Night a.m.	During Night a.m.	0 11	1.2	1 2	SE			✓	
10	31	21	21	During Night a.m.	2 45 pm	0 62	6.1	7.0	NE			✓	Sleet changing to snow, early a.m.
11	22	2	21			0		5 5	NW	✓			Light fog
12	26	3	16			0		4 2	NW	✓			
13	31	3	28	During Night a.m.		0		3 0	NE		✓		Dense fog
14	52	2	47		During Night a.m.	T		1 5	NE		✓		
15	55	20	35	During Night p.m.	During Night p.m.	T					✓		
16	70	25	31			0			S		✓		Solar halo
17	32	21	22	2 45 pm		T			N			✓	
18	22	16	18			0			NW			✓	High NW winds
19	18	10	16			0			NE			✓	
20	25	10	21	8 20 pm		0			NE			✓	
21	21	12	20			T	T		NW		✓		
22	36	6	30			0			S		✓		
23	35	10	21			0			SW		✓		
24	34	15	24	11 a m		0 08			E			✓	
25	34	15	20	12 15 pm		0 46			N			✓	
26	23	10	12		5 30 pm	0 12	1 4		NW			✓	Lunar halo
27	12	2	3	4 pm	6 20 pm	T	T		NW		✓		Solar halo, 10 a m
28	4	-5	2			T			NW	✓			Aurora 8 to 11 p m
29	3	-5	2			0			NW	✓			
30	11	-9	3			0			N	✓			
31													

SEE COVER FOR INSTRUCTIONS

John Roe, Cooperative Observer
Graceville, New York, Address

MONTHLY SUMMARY (TO BE FILLED IN BY OBSERVER)
DATES OF —
HIGH WINDS 18th.
GLAZE (FREEZING RAIN) ____
SLEET (SMALL ICE PELLETS) 10th.
HAIL ____
KILLING FROST 2nd
THUNDERSTORMS ____
FOG, LIGHT 11th
DENSE 13th
DEPTH OF FROZEN GROUND AT END OF MO ____ INS.

(NOT TO BE FILLED IN BY OBSERVER)
TEMPERATURE
MEAN MAXIMUM ____
MEAN MINIMUM ____
MEAN MONTHLY ____ DEPARTURE ____
MAXIMUM ____ DATE ____
MINIMUM ____ DATE ____
PRECIPITATION
TOTAL ____ INS DEPARTURE ____
GREATEST AMOUNT
IN ANY OBSERVATIONAL DAY: AMOUNT, DATE; COMPUTED FOR 24-HOUR PERIOD: AMOUNT, DATE
SNOW
TOTAL SNOWFALL ____ INCHES
GREATEST IN ANY OBSERVATIONAL DAY ____ DATE ____
NUMBER OF DAYS
MAX TEMP 90° OR ABOVE ____
MAX TEMP 32° OR BELOW ____
MIN TEMP 32° OR BELOW ____
MIN TEMP 0° OR BELOW ____
WITH 0 01 INCH OR MORE PRECIP ____
0 25 INCH OR MORE ____, 1.00 INCH OR MORE ____
CLEAR ____ PARTLY CLOUDY ____ CLOUDY ____

REMARKS BY OBSERVER
STATION ____
MONTH ____ 19__

FIG. 3. Form of the register of the cooperative stations of the U. S. Weather Bureau

The greatest difference between the two forms lies in the fact that the international form is arranged for three observations a day, while the U. S. coöperative observer makes one observation a day. Two columns are reserved for the extremes of temperature. A third column of this section contains numbers which are not of interest here. A great advantage, not provided in the international form, is the columns which indicate the time of the beginning and end of precipitation. Fundamentally important differences are noted in the case of cloudiness and wind: here the principle of making observations of an instantaneous state is not maintained, estimates of an average state during the day preceding the observation being given instead.

The remarking of differences does not imply, of course, a criticism; a scientific organizer is free to arrange the methods of observation so that he gets the greatest number of features characteristic of weather and climate. The student who builds up his climatography on the registers should be conscious of these facts, even if he is not compelled to go back to these fundamental sources.

It might happen, for instance, that the region whose climate has to be described is crossed by the borderline between two climatological networks. In the first of these, the average direction of the wind during the day is estimated; in the second, momentary wind directions are observed three times a day. It would be a source of errors if the different methods of observation escaped notice.

Usually, a space in the form is reserved for some simple statistical values which can be easily derived from the original observations.

I. 3. Some Remarks About Methods of Climatological Observations

The following discussion is not intended as an instruction on the arrangement of climatological observations. It serves only to emphasize difficulties of comparison which may arise with different elements.

I. 3. a. *Discontinuities in Climatological Series*

When the variations with time of an element at a place have to be investigated, its values at different times are compared. Changing the method of observation can cause a break, a discontinuity of the series. If a thermometer of the ordinary type is read over a period of ten years and is then replaced by a sling-thermometer, the temperatures during daytime become lower and night temperatures become slightly higher. Thus a discontinuity is caused, so that the second series cannot be compared with the first. The paradox is that while the sling-thermometer means a real improvement, this improvement destroys the continuity of the series. If after the introduction of the new procedure parallel observations are made in the old and in the new manner over a sufficiently long period, the values of the old series can be reduced to those of the new one. Suitable methods for such reduction are given in the following chapters.

An example of a nearly ideal "secular" station is represented by the Blue Hill Meteorological Observatory, south of Boston, Mass., founded by Abbott L. Rotch in 1885 Here observations have been made with suitable devices according to well-considered methods from the start, fifty-eight years ago. While numerous more modern methods were being introduced, the old methods of observing the most fundamental elements were continued simultaneously with the new ones. No change has taken place in the surroundings.

I. 3. b *Observing Air Temperature. Errors Caused by the Influence of Radiation and by Layering of Air Close to the Ground*

Observing the air temperature in order to make comparisons involves a considerable number of difficulties. Only one example need be mentioned here, which may contribute to a better understanding of the later sections of this book. A well-made thermometer shows the "true temperature of the air" if its bulb is in full heat-conductivity balance with the passing air. This obviously simple demand can hardly be fulfilled in free air because of the radiation from sun and sky, from the ground and from surrounding objects. The invention of the "aspiration-thermometer" eliminated the errors caused by radiation, for meteorological purposes. Because of some difficulties which need not be discussed here, climatological air temperature is defined by the indications of a well-constructed aspirated thermometer at the level of about 5 feet above the ground. Thermometers sheltered in a suitably placed "Stevenson screen" show temperatures which, generally, equal the indications of an aspiration-thermometer, so that the difference can be practically neglected. Thus, but for the existence of another difficulty, the problem of observing the climatological temperature would have been solved.

In the Pamirs (Central Asia, about lat. 38°N), in summer, a difference of about 60 F° [8] was observed between the temperature of the layer adjacent to the ground, and that of the $5\frac{1}{2}$ ft level above it.[9] Once in winter, in a depression in the eastern Alps, a

---

[8] The author gladly introduces a suggestion of Charles F. Brooks to discriminate in printing between °F, °C = an actual temperature and F°, C° = Fahrenheit, Centigrade degrees representing a temperature difference

[9] V Conrad, "Die klimatologischen Elemente und ihre Abhängigkeit von terrestrischen Einflüssen," Köppen-Geiger, *Handbuch der Klimatologie*, IB, p. 72.

temperature of −20°F was observed on the ground and, simultaneously a temperature of +30°F at the 180 ft level above the ground.[10]

The physical explanations of these conditions, interesting as they are, need not be set forth.   Here, only the fact counts that thermal stratifications of enormous intensity exist within thin layers above the ground.   Such abnormalities, though not of this intensity, are frequent everywhere.   We know this from an investigation in an exceedingly equable climate, near London, England.[11] From this study we also learn that the average vertical temperature gradient in Kew (England) is so great that the temperature decreases 1 F° per 3½ inches, within a layer from 1 ft to 4 ft above the ground, in June.   This is a gradient 625 times greater than the dry-adiabatic gradient.

The stratification of the air according to its specific gravity is, however, a very common climatic phenomenon in the concavities of the surface, when the outgoing radiation exceeds the incoming. The stratification is observed for a concave surface of any dimension.   It appears in the broad and narrow valleys of the high mountains of the earth, and is experienced in every furrow of the field.   Thus the layering of the air is of interest, not only for general considerations in observing air temperatures, but also as one of the most important features from which "Microclimatology," and especially "Orographic Microclimatology," makes its start.[12]

And now a problem arises concerning the temperature-observation: namely, to which layer should the observation of temperature be related, if the temperature varies in the lowest layers for every 3½ inches in the vertical direction as much as in the horizontal for about one degree of latitude.   The observations with the aspiration-thermometer, especially, become very inexact and ill-defined. Temperatures vary several degrees, according to the direction from which the air is sucked up.

There is only one expedient.   The place of observation at a climatological station should be a spot with good natural ventilation, open to all wind directions, so that thermal stratification is rare.   Furthermore, the thermometer-bulb should not be too close to the ground.   The International Meteorological Organization

[10] *Ibid* , p. 183.

[11] A. C. Best, *Transfer of Heat and Momentum in the Lowest Layer of the Atmosphere* (Geophysical Memoirs of the Meteorological Office, London, No 65, 1925).

[12] R Geiger, *Das Klima der bodennahen Luftschicht* (Braunschweig, 1927). See especially the second part "Orographische Mikroklimatologie," and Figure 30, p. 98.

advises fixing the thermometers at a height between 5 and 6½ feet. The thermometers in the screen of coöperative observers of the U. S. Weather Bureau are at about the height mentioned, and thus exclude the bottom layers with intense, vertical temperature gradients positive or negative.

We learn from this example that the exposure of the thermometer shelter must be chosen so as to avoid, if possible, local influences of radiation and thermal stratification. Generally, only the former influence is mentioned, but the latter is also a great disadvantage. Both factors seriously impair the comparableness of the data. The thermometer shelter should be located so that the recorded temperatures are representative of the largest possible surrounding area, a condition which also makes for the most comparable results. It is obvious how badly the continuity of a series of temperatures may be disturbed by shifting of the shelter.

I. 3. c. *Measuring Precipitation*

Measuring precipitation offers a difficult problem. None of the various methods gives entirely satisfactory results. The amount of water or snow collected in the gauge depends, to a certain degree, upon the wind velocity and the air resistance of the particles of precipitation. The stronger the wind and the greater the air resistance to the particles, the smaller is the amount caught in the gauge compared with that collected under otherwise similar conditions during absolute calm. Shielded gauges, invented by F. E. Nipher, give incomparably better results [13]

In general, the normal climatological stations are not yet equipped with shielded gauges, so that the wind and the different types of precipitating particles still influence the measurements. hence, a degree of comparableness can be achieved only by using identical gauges installed in a uniform way.

This is true for the United States, half a huge continent, where the exposure of rain-gauges follows sound uniform principles. If some reader is interested, however, in a study of precipitation in Europe, he is confronted with much more complex conditions.

[13] For further information, see· C. F. Brooks, "Windshield for Precipitation Gauges" (*Transactions of the American Geophysical Union*, 1938, p 539), C F Brooks, "Further Experience with Shielded Precipitation Gauges on Blue Hill and Mount Washington" (*ibid*, 1940, pt II, p 482), H S Riesbol, "Report on an Exploratory Study of Rain Gauge Shields and Enclosures at Coshocton, Ohio" (*ibid*, p 474), F. N. Denison, "A Report on the Difference Between the Precipitation Records as Taken on the Standard Canadian and U S Rain-Gauges" (*Bulletin of the American Meteorological Society*, February 1941, p 65)

From the following table, the areas of the receivers of types of rain-gauges used in some countries in Europe may be seen, and for comparison, the area of that used by the U. S. Weather Bureau:

Austria	78 sq. inches	
England	31 sq. inches	type I
England	20 sq inches	type II
France	62 sq. inches	
Germany	31 sq. inches	
U. S. Weather Bureau	50 sq. inches	

Perhaps the size of the receiver has no great influence upon the results. In any case, these numbers show that there is no uniformity.

As was said above, measurements of precipitation depend on wind velocity. Because this varies enormously with the height above the ground, it is clearly a scientific requirement that the rim of the receiver should be at the same height in all countries. In reality, the heights are:

Austria	59 inches
England	12 inches
France	59 inches
Germany	39 inches
U. S. Weather Bureau	34 inches

This is a violation of the scientific requisite of uniformity. The climatologist who has the task of giving the distribution of precipitation over a region should be aware of these inequalities and discontinuities.

CHAPTER II

GENERAL STATISTICAL CHARACTERISTICS OF CLIMATOLOGICAL ELEMENTS

THE INTERNATIONAL MODEL of the meteorological (climato-logical) register, covering the period of a month, contains about 750 observations; 9,000 observations for the year. The registers of a small climatological network with one hundred stations yield 900,000 observations in one year. The 5,500 co-operative stations of the U S. Weather Bureau produce registers with about 12 million entries per year. The human brain is not capable of comprehending such a huge mass of numbers. Thus, a statistical treatment of observational data is necessary.

II. 1. THE SYSTEM OF CHARACTERISTICS

The following pages present a selection of such methods of mathematical statistics as are suitable for abstracting the common characteristics of climatological elements. L. W. Pollak has made a simple survey of these characteristics, and his system is based on the amount of mathematical operations necessary to compute them. They are divided into groups: [1]

1) Primitive characteristics
2) Elementary characteristics
3) Higher characteristics.

II. 2. PRIMITIVE CHARACTERISTICS

Primitive characteristics can be derived from the original ob-servations by direct copying, by grouping under special headings, by counting, and by similar primitive methods. Examples will be given to pave the way to better understanding and to illustrate the close connection with climatological topics.

II. 3. THE VARIATE. THE ABSOLUTE EXTREMES. THE RANGE OF VARIATION

An example is the January minimum temperature at Mount Washington, N. H., for the four years 1938 to 1941.

[1] L. W. Pollak, Charakteristiken der Luftdruck-Frequenz-Kurven und verallgemein-erte Isobaren, *Prager Geophysikalische Studien*, vol I (Prague, 1927)

First, a list of the data, as in Table 1, is required. In this list, the extremes of the minima are easily checked for each year. It is practicable to italicize the maxima, and to mark the minima by asterisks. Then the extremes of the entire sample are seen at first

TABLE 1   DAILY MINIMA OF TEMPERATURE (°F) AT MOUNT WASHINGTON, N. H., FOR FOUR MONTHS OF JANUARY

(*Mount Washington Observatory News Bulletin*)

January	1938	1939	1940	1941	
1...	2	− 9	−14	6	
2..	5	−10	−14	9	
3. .	1	−21	−10	9	
4	8	2	−11	6	
5	0	16	− 3	− 2	
6	1	19	−14	− 2	
7	15	*20*	− 6	1	
8	− 3	− 2	0	0	
9	−13	0	0	6	
10.	−10	16	0	− 3	
11	−10	2	0	7	
12	1	−13	5	− 2	
13.	4	−12	*11*	−30*	
14	− 4	− 4	8	−30*	
15	− 4	− 8	4	−24	
16	− 5	6	−30*	− 1	
17	− 5	−15	−30*	19	
18	−15*	−15	−28	*21*	
19	− 2	0	−16	−12	
20.	2	− 4	−15	−12	
21	4	− 4	−11	− 6	
22.	8	2	−19	12	
23	9	−36*	− 8	− 6	
24	11	−10	0	− 4	
25	*19*	−22	− 9	3	
26	− 5	−29	−14	− 5	
27	−13	−13	−12	6	
28	−11	−12	−13	5	
29	−13	8	− 8	−12	
30	− 4	4	− 8	− 1	
31 .	− 6	− 1	− 5	− 4	
$\overline{m}$	− 1 1	− 4 7	− 8 7	− 1 5	−4.0

glance. These are the *absolute extremes* of the sample of the variate, that is the variable quantity. In Table 1, the variate is repre-sented by the daily minimum temperatures for the months of

January.   We get:

$$\begin{array}{ll} \text{Absolute maximum} = & 21°F \\ \text{Absolute minimum} = & -36°F \\ \hline \text{Difference} \qquad = & 57\ F° \end{array}$$

This difference is called the *range of variation*, and defines the entire range of the sample.   The range is divided into a number of equal *class intervals*.

II. 4. CLASS INTERVALS.   FREQUENCY DISTRIBUTION. ABSOLUTE AND RELATIVE FREQUENCY

The choice of the size of the class intervals is dependent upon the number of observations included in the sample and on the degree of exactness required.   For the conditions of the model in Table 1, 4F°-intervals are more or less suitable, because the sample consists of only 124 elements.

TABLE 2.   FREQUENCY TABLE.   GROUPING THE DATA INTO CLASS INTERVALS OF 4 F°

Number (*) (1)	Class Limits (°F) (2)	Class Center (3)	Frequency Absolute (4)	Frequency Percentage (%) (5)
1 .	−36 to −33	−34 5	1	0 8
2	−32 to −29	−30 5	5	4 0
3.	−28 to −25	−26 5	1	0 8
4	−24 to −21	−22 5	3	2 4
5	−20 to −17	−18 5	1	0 8
6	−16 to −13	−14 5	15	12.1
7	−12 to − 9	−10.5	16	12 9
8	− 8 to − 5	− 6 5	13	10 5
9	− 4 to − 1	− 2 5	19	15 3
10	0 to 3	1 5	19	15.4
11	4 to 7	5 5	13	10.5
12	8 to 11	9 5	9	7 3
13	12 to 15	13 5	2	1 6
14	16 to 19	17 5	5	4 0
15	20 to 23	21 5	2	1 6

* For some statistical purposes it is more desirable to begin with no. 0 (zero).   See: E. Czuber, *Die statistischen Forschungsmethoden* (Wien, 1921), p. 91, and L W. Pollak, *Prager Geophysikalische Studien*, vol I, pp 33 and 34

These class intervals are seen in the second column of Table 2. In many cases it is more advantageous to give the *central height* or midpoint of the class interval, as in column 3.   The first column contains the serial numbers of the class intervals.

The number of observations falling within the limits of the consecutive intervals is counted.   The resulting series represents the *frequency distribution* of the sample in question.   Column 4 in Table 2 gives the frequency distribution of minimum temperatures, etc.   These numbers are called *absolute frequencies*.   For fundamental climatological purposes of comparison, *relative frequencies* are more expedient; these are absolute frequencies expressed in per cent of the total number (n) of observations.   (In the example of Table 2, $n = 124$.)   Examples of the great usefulness of frequency distributions, which often mean an indispensable supplement to the usual averages, can easily be given.

The figures in Table 3 are taken from a war climatology (World War I), when it was necessary to give an effective comparison of summer temperatures in Central Europe (Vienna, Austria: 48.2°N, 16.3°E, 660 feet) and the southern Adriatic coast (Scutari. 42.1°N, 19.4°E, 70 feet).   One might think that a simple comparison of temperature maxima would be illustrative.   In reality, in Scutari, a maximum of 98°F has been observed within a period of more than 30 years, in Vienna 98°F was also observed, though within a period of 70 years.   The maxima are practically identical.

A frequency distribution of temperature in Vienna and Scutari, however, shows immediately the characteristic and great climatic differences of the two places in spite of the fact that the comparison considers the temperatures only at 7 A.M. of the summer months of a synchronous period of 7 years.   Each of the samples in Table 3 comprises $92 \times 7 = 644$ observations.

TABLE 3   DISTRIBUTION OF RELATIVE FREQUENCIES (%) OF TEMPERATURE AT 7 A M AT A PLACE IN CENTRAL EUROPE (V) AND AT A PLACE ON THE SOUTHERN ADRIATIC COAST (S) IN SUMMER
(JUNE, JULY, AUGUST)

Class interval F°	V %	S %
41 0 to 49.9.	0 5	—
50 0 to 58 9	35 4	—
59 0 to 67 9	59.2	19 9
68 0 to 76 9	4.9	70 3
77 0 to 85 9	—	9 8
Σ	100 0	100 0

From Table 3 it is seen that at 7 A.M., after the cooling during the night, 95% of all observations are below 68°F in Central Europe; on the Adriatic coast 80% exceed 68°F.   At first glance, it is also seen that the distribution is less scattered in Scutari (3 classes) and more in Vienna (4 classes).

These facts reveal a drastic contrast of temperature conditions in the two regions. It is not the absolute height of the maxima which counts. It is, rather, the fact that the greatest frequency is shifted to the next higher class on the Adriatic Sea. That means great frequency of high temperatures during the night and in the morning hours, a condition enervating for people accustomed to a more moderate climate. The frequency distribution points to the greater monotony of high temperatures in the regions of the Adriatic Sea during the summer.

The foregoing example shows the value of relative frequencies for purposes of comparison. Besides, the size of the class interval is notable. With the sample in Table 3 the size of the interval is not less than 9 F°, sufficient for the desired purpose.

II. 5. HISTOGRAM AND FREQUENCY CURVE. THE MODE. THE MODAL CLASS

As a consequence of the relatively small intervals in Table 2, the distribution takes a rather irregular shape, which is well illustrated by the histogram of Figure 4. This figure contains two methods of representation: the histogram, "rectangular areas standing on each grouping interval showing the frequency of observations in that interval,"[2] and the frequency curve which has been smoothed. In reality, the variation in question is discontinuous, because the variate is confined to whole degrees Fahrenheit (see Table 2). The frequency curve is more usual in climatological statistics, and there is no real reason "for insisting on the histogram form." The latter may be more appropriate from a theoretical standpoint. The continuous curve gives a better survey, and the maximum of the curve indicates the approximate value of the *mode*, the most frequent value. This procedure is based on the assumption that a large number of observations are available to give a smooth distribution of frequencies. In the present example, that is not realized.

[2] R A Fisher, *Statistical Methods for Research Workers* (6th ed , London, 1936), p 38 The frequencies can be plotted on the ordinates through the midpoints of the class intervals "The tops of all the ordinates are joined by a broken line," i e the *original* frequency curve which appears as a *smoothed* line in Figure 4. *See* D Brunt, *The Combination of Observations* (Cambridge, England, 1931), p. 7.

Another important concept is the *modal class*, that is, the class interval into which the greatest number of observations fall.

In the example of Table 2, there are two equal *modal class intervals*, nos. 9 and 10. This offers a certain difficulty in com-

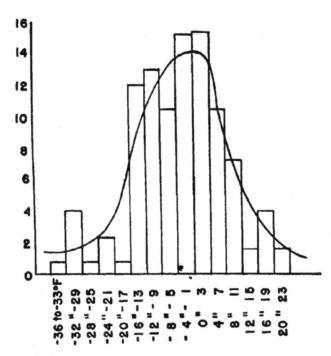

FIG. 4. Histogram of relative frequencies A smoothed frequency curve is added. Minimum temperatures of four months of January at Mount Washington, N. H

puting the temperature which represents the mode. Even in this case, histogram and frequency curve permit an estimate of the mode. In the example in question, the mode $= (-2.5 + 1.5)/2 = -0.5°F.$

Generally, the value of the mode should be approximated by a simple linear interpolation; the equation reads:[3]

$$\text{Mode} = L_{mo} + \frac{f_a}{f_a + f_b}\ C$$

[3] H. Arkin, R R. Colton, *An Outline of Statistical Methods* (4th ed , New York, 1939), p. 23

Where

L_{mo} = Lower limit of modal class
f_a = Frequency of class interval above modal group
f_b = Frequency of class interval below modal group
C = Size of class interval

As stated before, in the example of Table 2 the class intervals nos. 9 and 10 include equal numbers of frequency. Therefore they are combined into one modal class with a frequency number = 38.

In order to avoid unequal intervals, nos. 11 and 12 are taken collectively as f_a and nos. 7 and 8 as f_b.

Thus we get from Table 2:

$$L_{mo} = -4°F \text{ (lower limit of interval no. 9)}$$
$$f_a = 13 + 9 = 22 \text{ (nos. 11 and 12)}$$
$$f_b = 13 + 16 = 29 \text{ (nos. 8 and 7)}$$
$$C = 8°F$$

Therefore.

$$\text{Mode} = -4 + \frac{22}{22 + 29} 8 = -0.55°F$$

Accidentally, the two results are in perfect agreement.

Generally, frequency distributions are necessary and indispensable in climatological investigations. This does not hold to the same extent where mode is considered. Variates taken from the realm of climatology, in particular, show remarkable frequency curves with two or more modes, so that the mode can no longer be considered as an unequivocal characteristic. Frequency curves of winter temperatures, for instance, show this phenomenon, especially at places located in a transition zone between an oceanic climate and a strongly continental one.

Two modes exhibit apparently just about the same frequency: one, at a temperature of some degrees below the freezing point, corresponding to the average temperature of fresh continental air masses (Leningrad, 59.9°N, 30 3°E. about 20°F); the other mode, at about the freezing point. This means that the heat content of the advected oceanic air masses at higher temperatures is expended in the process of melting the snow-cover, its thermostatic effect.

We learn from this example that, in the great number of similar cases, two (and even more) modes can exist, so that the bulk of observations cannot be characterized by one number alone, in contradistinction to the arithmetical mean. In other respects, also, the mode of climatic series happens to have properties which do not make it a desirable statistical element. On the other hand, one sees from the last example that the double mode of the temperature at a place (like Leningrad on the Baltic Sea) leads to a clear physical analysis of an important climatic feature. Even in this respect, it should not be forgotten that the frequency curve conveys the same knowledge, and moreover a comprehensive view of all observations.

II. 6. ELEMENTARY CHARACTERISTICS.
MEDIAN, QUARTILES, DECILES

The method of computing the median stands between the primitive and the elementary characteristics, because there are—at least theoretically—two methods, the first of which leads immediately to a primitive element.

All items of which the sample consists are arranged according to size. This series is called the *array*. If the number of these items is odd, then the middle value is called the *median*. If the number is even, then the average of the two central items is defined as the median. This method obviously represents the median as a primitive characteristic.

Example: Take the series of minimum temperatures in January 1938 from Table 1 and arrange them according to magnitude. The series of Table 4 A results. The median $= -2°F$. If the last

TABLE 4. A EXAMPLE FOR AN ARRAY. (TEMPERATURE MINIMA JANUARY 1938, FROM TABLE 1)

No	1	2	3	4	5	6	7	8	9	10	11
	−15	−13	−13	−13	−11	−10	−10	−6	−5	−5	−5

No.	12	13	14	15	16 (median)	17	18	19	20
	−4	−4	−4	−3	−2	0	1	1	1

No	21	22	23	24	25	26	27	28	29	30	31
	2	2	4	4	5	8	8	9	11	15	19

B GROUPED DATA

No	°F	Frequency	$\sum_{1}^{4} F_i$
1	−15 to −7	7	7
2	− 6 to 2	15	22
3	3 to 11	7	29
4	12 to 20	2	31

item (no. 31 = 19) were omitted so that the series had an even number of items = 30, the median would be the average of the two central items, nos. 15 and 16:

$$\frac{(-3) + (-2)}{2} = -2.5°F$$

This method of arranging is "primitive," but as a rule very tedious, because of the magnitude of climatological samples.

The second method must be applied to samples which are already grouped into class intervals. For this purpose, a sample with a greater number of items is necessary. Table 5 contains a sample

TABLE 5 FREQUENCY DISTRIBUTION OF MINIMUM
TEMPERATURE (°C) OF 25 MONTHS OF JULY[*]

(Coordinates for the place K 48 1°N, 14 1°E, 1280 feet)

(1) No. (i)	(2) Central height °C	(3) F_i	(4) $\sum_{n=1}^{n-15} F_i$
1	6 45	1	1
2	7 45	1	2
3	8 45	4	6
4	9 45	15	21
5	10 45	52	73
6	11 45	84	157
7	12.45	131	288
8	13 45	121	409
9	14 45	108	517
10	15.45	114	631
11	16.45	75	706
12	17 45	45	751
13	18 45	13	764
14	19.45	9	773
15	20.45	2	775

[*] Numbers of Table 5 taken from F. Steinhauser, *Meteorologische Zeitschrift*, 1935, p 207.

of daily minimum temperatures (°C) for 25 months (of July at a place in the foothills of the eastern Alps).

The absolute extremes of this sample are: 6.3°C and 20.9°C
The range of variation is. 14.6 C°
The number of items is· 775
As this is a sample of greater magnitude with an exactness of

0.1 C° of the individual observations, 1 C° is chosen as a suitable size for the class intervals.

The first interval reaches from 6.0°C to 6.9°C
the second from 7.0°C to 7.9°C
the last from 20.0°C to 20.9°C

The central heights of the class intervals are therefore:

$$\frac{(6.0 + 6.9)}{2} = 6.45°C \text{ etc.}$$

The first column of Table 5 contains the number of the class interval; the second, the central height of the class interval; the third, the frequency within the class intervals; and the fourth, the series of cumulated frequencies.

This series $\sum_{i=1}^{i=15} F_i$ is calculated as follows· The frequency of the interval no. 1 = 1; the frequency of nos. 1 and 2 = 1 + 1 = 2 etc. The last sum (at no. 15) represents therefore the whole size of the sample, i.e., 775.[4]

The median value is defined in such a way that 50% of the sample lies below it and 50% above. The whole size is:

$$N = 775$$
$$50\% \text{ of } N: \quad \frac{N}{2} = 387.5$$

From column 4, it is seen that this limit is surpassed within the 8th class interval. Therefore the median is located between 13.0°C and 13.9°C. In this case, a simple linear interpolation leads also to the result.

The 8th interval contains 121 items, of which 99.5 (387.5 − 288) should lie below the median value. Consequently, the interval of 1.0°C (if we start from 12.9°C) has to be divided in the propor-

---

[4] Σ, the capital Greek S, means the sum of a given number of consecutive elements of a series.

$\sum_{a}^{b}$ means that the sum is to be taken from the element with the number (subscript) a to an element no b

tion of

$$\frac{99.5}{121} = 0.82$$

so that the median is at $12.9°C + 1.0 \times 0.82 = 13.72°C$.[5]

According to the same principle, the entire size of the sample can be divided into an arbitrary number of equal parts, which are symmetrical about the median. It is usual to divide the sample into 10, or into 4, parts. The limits of the first are called *deciles*, those of the second, *quartiles*.

In the example of Table 5, the limits of the deciles are given by the cumulative sums of $N/10 = 77.5$, with the following results:

Decile no.	1	2	3	4	5
Upper limit number of items	77 5	155 0	232 5	310 0	387.5
Decile no.	6	7	8	9	10
Upper limit number of items	465 0	542 5	620 0	697 5	775 0

The procedure of computation is now identical with that used for the median. The first decile contains 77.5 items, and therefore it lies within the 6th class interval between $11.0°C$ and $11.9°C$. Because the first 5 class intervals contain 73 items, the 6th class interval participates with 4 5 items to give the first decile. Therefore, the entire size of the 6th class interval has to be divided in the proportion of

$$\frac{4.5}{84} = 0.05$$

The upper limit of the first decile is consequently at:

$$10\ 90 + 1.0 \times 0.05 = 10.95°C$$

Herewith it is assumed that the sample is composed of discrete numbers, e.g., 6.0, 6.1, etc.

[5] Part B of Table 4 shows the data of part A grouped into class intervals of 9F° There

$$\frac{N}{2} = 15\ 5$$

Because the first interval contains 7 items, 8.5 items (15 5 − 7 = 8.5) of the second interval lie below the median Therefore this interval of 9 F° with 15 items must be divided in the proportion of

$$\frac{8\ 5}{15} = 0.57$$

and the median lies at: $-7° + 9° \times 0\ 57 = -2°F$. It is worth noting how good the agreement is between the method of an array (see Table 4A: − 2°F) and the method in question, even for a sample of only 31 elements.

Reduced to one decimal place, we get

$$11.0°C$$

The second and first deciles together contain 155.0 items. The second decile also lies within the 6th class interval which participates with 82 items. The class interval must be divided in the proportion of:

$$\frac{82}{84} = 0.98$$

And the upper temperature of the second decile lies at:

$$10.9 + 1.0 \times 1.0 = 11.9°C \text{ etc.}$$

Thus the following division of the sample results:

	°C
Absolute minimum	6 3
Decile 1	11 0
Decile 2	11 9
Decile 3	12 5
Decile 4	13 1
Decile 5	13.7 = median
Decile 6	·14 4
Decile 7	15 1
Decile 8	15 8
Decile 9	16.8
Absolute maximum	20.9

The deciles divide the sample into equal parts of 10%; the quartiles into parts of 25%. As stated above, there is no other difference in principle between them. The computation of the quartiles needs no further explanation.

In the example of Table 5, $N = 775$, the cumulative sums for the quartiles would be 193.75; 387.50; 581.25; 775.00. Therefore:

Absolute minimum	Lower quartile	Median	Upper quartile	Absolute maximum
6 3	12 2	13.7	15.5	20.9°C
	5 9 C°		1 5 C°		1 8 C°		5 4 C°	

It is characteristic of the element under investigation that 50% of all observations lies within 3.3 C° (between 12.2°C and 15.5°C). The other 50% is distributed over a range of 11.3°C altogether. These facts give a desirable picture of the rarity of greater deviations from the mean value.

The size of the interval between upper and lower quartiles is always of interest because 50% of the entire variate falls within these limits and 50% beyond them.   There is the equal *likelihood* that an observation falls inside or outside of the limits of the quartiles.   The small range between the lower and upper quartiles shows that the density of the items is much greater inside this space than outside.

The three characteristics of the frequency curve—*median, quartiles, deciles*—have been discussed, and a simple method has been shown of computing these statistical values.   This has been necessary because in the English meteorological and climatological literature these characteristics are frequently used.   Furthermore, these methods assume their real value only if the size of the sample is so large that half the number of its items represents a series of many hundreds.   If, then, the comparison of the characteristics in question showed significant differences between two or more periods of years, we should infer "that the climate (or the methods of measuring it) had materially altered." [6]

On the other hand, incontestable climatic series of great length —large samples—are rare; with small samples, the results from these characteristics are often disappointing.   Further discussion of this subject is therefore superfluous.   The frequency curve itself is undoubtedly one of the best and finest means in the hand of the climatologist.

[6] R A Fisher, *Statistical Methods*, p 42.

CHAPTER III

SPECIAL STATISTICAL CHARACTERISTICS OF CLIMATOLOGICAL SERIES

III. 1. The Arithmetical Mean

IT GOES without saying that the arithmetical mean is defined by:

$$\bar{a} = \frac{1}{n} \sum_{\imath=1}^{\imath=n} a_\imath$$

where $\bar{a}$ denotes the arithmetical mean, $a_\imath$ represents any number from a_1 to a_n and n is the number of observations which constitute the series.

The computation of the arithmetical mean even of large series is simple, especially if a calculating machine is at hand. If the size of the sample is extraordinarily large, there are methods for computing the arithmetical mean from the sample divided into class intervals, when the midpoints and the frequencies are known. For these, the reader is referred to textbooks of statistical methods, since such cases are not sufficiently significant for climatology.

The use of the arithmetical mean (average) in climatology is common and need not be fully discussed. Here, however, a few remarks about averages of extremes may be opportune, because of frequent misunderstandings. The following examples are taken from Table 1, minimum January temperatures on Mount Washington, New Hampshire.

(a) Mean daily extremes; in this case "minima"
1938: -1.1; 1939: -4.7; 1940. -8.7; 1941: $-1.5°$F

(b) The x-years' average of the mean daily extremes; in this case the four-years' average of the mean daily minima $= -4°$F

(c) Average absolute extremes calculated from the absolute extremes of each year (month etc.); in this case
lowest minima: $(-15, -36, -30, -30):4 = -28°$F
highest minima: $(+19, +20, +11, +21).4 = +18°$F

(d) The x-years' average range of the extremes, in this case of the minima, is 46 F°

The examples are taken arbitrarily from the topic of temperature in order to dispense with additional tables; the underlying concepts are applicable to any element of scalar character.

III. 2. DEVIATIONS. AVERAGE VARIABILITY.

STANDARD DEVIATION. NORMAL DISTRIBUTION

There may be a series of n numbers representing the results of n observations:

$$a_1, a_2, \cdot \cdot a_n$$

The arithmetical mean is:

$$\bar{a} = \frac{1}{n} \sum_{i=1}^{i=n} a_i$$

Then the differences between the single elements of the series a_i and the arithmetical mean, i.e.,

$$d_1 = a_1 - \bar{a}; \qquad d_2 = a_2 - \bar{a} \cdot \cdot d_n = a_n - \bar{a}$$

are called deviations, which play the greatest role in the representation of statistical and also of climatological series of numbers.

The sum of the deviations from the arithmetical mean is zero, if the signs of the deviations are considered.[1]

The sum of the deviations, without regard to sign, divided by their number is called *mean deviation* (μ) in statistical mathematics and *average variability* (*AV*) in climatology. Therefore:

$$\mu = \frac{\Sigma |d|}{n}$$

Here: $\Sigma | \ |$ means the sum of the absolute values, ignoring signs:

$$\mu = (|d_1| + |d_2| + \cdots + |d_n|) : n$$

Table 6 gives examples for the computation of deviations and of the average variability (mean deviation). The dates of the *last killing frost* in Vancouver, B. C., and in Bismarck, N. D., in the

---

[1] In American papers and books the expression "deviation" is frequently replaced by the term "departure" and "mean deviation" by "mean departure" The term "departure" is seemingly not so common in British investigations and is even lacking in the *Meteorological Glossary* (London, 1939), and in Shaw's *Manual of Meteorology* The term "average variability" is usual in numerous climatological papers of many countries and has its advantages from the didactic standpoint

TABLE 6   DATES OF THE LAST KILLING FROST IN VANCOUVER, B.C., AND IN BISMARCK, N.D , AND DEVIATIONS (d) FROM THE MEAN DATE (ṁ)

Year	Vancouver, B C				Bismarck, N D			
	Date (1)	Year day* (2)	d (3)	d² (4)	Date (5)	Year day* (6)	d (7)	d² (8)
1901	IV 24	114	+ 6	36	VI 7	158	+24	576
2	12	102	− 6	36	IV 29	119	−15	225
3	15	105	− 3	9	V 5	125	− 9	81
4	23	113	+ 5	25	14	134	0	0
5	III 31	90	−18	324	12	132	− 2	4
6	IV 12	102	− 6	36	27	147	+13	169
7	29	119	+11	121	14	134	0	0
8	28	118	+10	100	2	122	−12	144
9	21	111	+ 3	9	13	133	− 1	1
10	14	104	− 4	16	17	137	+ 3	9
			+35				+40	
Σ		1078	−37	712		1341	−39	1209
			72				79	
ṁ	IV 18	108	7 2		V 14	134	7.9	
σ				8 44				11 00

* Number of the day from first of January.
    The data are taken from Frank J Bavendick, "Climate of North Dakota," p 1048, and Lawrence C. Fisher, "Climate of Washington," p 1175, *Climate and Man*, Yearbook of Agriculture (Washington, D. C , 1941).

ten years 1901 to 1910 are given. For computation, the dates (columns 1 and 5) have to be converted to numbers of days starting January 1. (February 1 = no. 32, etc.; see below, Appendix IV) These numbers are shown in columns 2 and 6. The average dates (arithmetical means) are the 108th and 134th days of the year, or April 18 and May 14 respectively. In columns 3 and 7, the deviations are shown and their sum total (disregarding the signs) is given. The sums of the positive and of the negative deviations should be identical (small differences being due to rounding off the arithmetical mean (ṁ) to integers).[2]

The sum of the absolute values of the deviations divided by their number (10) equals the average variability, which represents one of the most valuable climatological elements derived from the observations. From the examples, the variability of the date of the last killing frost is somewhat greater in the interior of the continent than at the coast. The principal contrast is given by the differ-

[2] The differences should be checked In the case of column 2, the exact ṁ = 107.8. Each of the 10 deviations is afflicted therefore with an error of ±0 2 Hence, the difference 10 × 0.2 = 2, in agreement with the result in column 3 In column 6 the exact mean is 134 1 instead of 134 The difference is 0.1 × 10 = 1.

ence in the average dates.   Inland, the average last killing frost occurs 26 days later than at the coast.

A rough measure of the variability is also given by the difference between the absolute extremes, i.e., 29 days at Vancouver and 39 days at Bismarck, N. D.   This measure is called "rough" because it depends upon only *single* values   For instance, if the series began with 1902, the differences would have been 29 days (unchanged) and 28 days, respectively.   The average variability considers each value of the series, and is the usual method of characterizing the scatter of the element in question.

The second method is to compute the *standard deviation* [3] (σ)

$$\sigma = \sqrt{\frac{\Sigma d^2}{n}}$$

which is the best and most exact measure of scattering.   In climatological investigations it is not so frequently used, since the calculation is more complicated.   This problem will be treated later.

The example of Table 6 gives the following results (columns 4 and 8):

$$\Sigma d^2{}_V = 712 \qquad\qquad \Sigma d^2{}_B = 1209$$

$$\sigma_V = \sqrt{\frac{712}{10}} = 8.44 \qquad \sigma_B = \sqrt{\frac{1209}{10}} = 11.00$$

The standard deviation, the scatter of the date of the last killing frost, is about 30% greater inland than at the coast.

From the frequency distribution shown in Figure 4 and Tables 4B and 5, one draws the simple and natural conclusion that extreme values are rarer than values close to or within the modal class interval.   Under certain conditions, this experience holds, too, in the case of *deviations* if the inferences are logically changed. This means that the smaller the deviations, the more frequent they are.

The ideal case is realized by the so-called *normal distribution*, or Gaussian distribution, which governs deviations independent of one another.

---

[3] This is the "standard deviation" of the single value; that of the arithmetical mean is:

$$\sigma' = \frac{\sigma}{\sqrt{n}} = \frac{\sqrt{\Sigma d^2}}{n}$$

There should be no effective reason why the deviations, if their number be large, may not take any value from $-\infty$ to $+\infty$. Then, the frequency curve is a bell-shaped curve which fulfills the equation:

$$y = \frac{1}{\sigma\sqrt{2\pi}} e^{-(d^2/2\sigma^2)}$$

where y means the frequency ordinate at the distance d (deviation) from the center of the curve (Fig. 5), $\pi = 3.142$ (Ludolf's number),

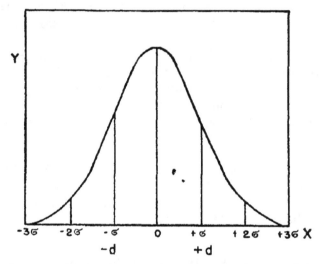

FIG. 5. Normal distribution showing the areas (per cent) between ordinates in distances of one, two, three Standard Deviations left and right from the center (frequency of the arithmetical mean)

$e = 2.718$ (base of the natural logarithms). This is the law of Karl Friedrich Gauss.[4]

Obviously, the curve represented by this equation is symmetrical, with the greatest frequency at the center, because equal values of y are related to $(-d)$ and to $(+d)$. The greatest fre-

---

[4] For a more correct representation, L H C. Tippett (*The Methods of Statistics*, London, 2nd ed., p. 54) may be quoted. "The frequency curve is deduced from a histogram and consequently *areas* under the curve, not heights of the ordinates, represent frequencies It is therefore appropriate to use the notation and ideas of the integral calculus, to imagine about any given value of x an elemental sub-range dx, and to regard the area under the curve between ordinates drawn at the limits of the sub-range as an element of frequency, df. Then this elemental strip may be regarded as a rectangle of height y and $df = \dfrac{N}{\sigma\sqrt{2\pi}} e^{-d^2/2\sigma^2} \cdot dx$," where N denotes the number of items

quency occurs at $d = 0$, if

$$(e^{-(d^2/2\sigma^2)})_{d=0} = 1$$

Therefore the arithmetical mean is the most frequent value, and coincides with the values of the median and the mode. This is valid only if there is a symmetrical distribution of frequency of the deviations with an absolute maximum at $d = 0$.

Gauss's equation shows that a normal frequency distribution is fully defined by its respective standard deviation σ.

Geometrically, σ is the distance on either side of the center where the slope is steepest, at the points of inflexion of the curve. (See Fig. 5.)

In practical applications we are not so much interested in the frequency at any distance from the center. It is much more important to know the total frequency (or probability) beyond this point (for instance beyond the distance σ in Fig. 5). This frequency or probability is represented by the area included between the ordinate at the chosen point, the tail of the curve, and the X-axis.

For deviations expressed in units of σ, (d/σ), tables of this total frequency, or probability integral, have been constructed, from which, for any value d/σ, we can find what fraction of the total sample has larger deviations

III. 3. A Practical Application of the Standard Deviation in Climatology

The probability (P) [5] that a deviation exceeds 1, 2, or 3 times the standard deviation is 0.317; 0.047; 0.003. The numbers illustrate the rapid decrease of the probability of increasing deviations.

In other words, the area between the ordinates at: $\pm\sigma$ (Fig. 5) $\pm 2\sigma$, $\pm 3\sigma$ includes:

between $-\sigma$ and $+\sigma$: 68.260%
between -2σ and $+2\sigma$: 95.346%
between -3σ and $+3\sigma$: 99.729%
between -4σ and $+4\sigma$ 99.994%

[5] That is the ratio of the number of favorable events to the total number of possible events. If a coin is tossed, only "heads" or "tails" can appear. The probability of heads (e g) is therefore 1/2.

In climatological investigations, the *probable error* (*f*) can be useful because it includes exactly 50% of the sample between $+f$ and $-f$. The probability that a deviation is $\geqq f$ is therefore 1/2.

The distance of the ordinate from the center which borders this area of the curve is called *quartile distance*, a term the meaning of which is clear from previous explanations. The percentages of the areas included between the ordinates at $\pm f$, $\pm 2f$, etc. are:

$$\begin{aligned}
&\text{between} + f \text{ and} - f \;\; 50\,00\% \\
&\text{between} +2f \text{ and} -2f\text{:}\; 82.26\% \\
&\text{between} +3f \text{ and} -3f\text{:}\; 95.70\% \\
&\text{between} +4f \text{ and} -4f\text{:}\; 99.30\%
\end{aligned}$$

The probable error is usually computed from the standard deviation according to the relation:

$$f = 0.6745\sigma$$

Example of the practical use of the division of the area of the curve of normal distribution

Often the climatologist has to determine the rareness of an actual observation of temperature, precipitation, etc., or else of a mean intensity, or amount, such as a mean of temperature, sum of precipitation of an individual month, etc. Statements to that effect are necessary on the occasion of an expert opinion and in the description of a climate. Questions may arise such as: Is a wind velocity of x mi/hour, is a temperature of x degrees below zero, is a downpour of x inches in an hour, etc., a normal event in a given climatic region, or is the observed value rare, or very rare? The courts are especially interested in such estimates because in the latter case the judge can declare the event to be a *force majeure* (superior force). For many other purposes, including military ones, it is highly important to know the limits between which normal values of the climatic elements can vary.

Usually such questions are answered by the climatologist according to common sense.

The division of the area of the curve of normal frequency distribution provides a transition from guessing to quantitative estimation. If a long series of observations of the element in question is at hand, one can calculate σ or f (standard deviation or probable error).

Then the following nomenclature can be used:

below: $-3\sigma(-3f)$:   extremely subnormal $= ES$
between: $-3\sigma(-3f)$ and $-2\sigma(-2f)$:   greatly subnormal $= GS$
              $-2\sigma(-2f)$ and $-\sigma(-f)$:   subnormal $= S$
              $-\sigma(-f)$ and $+\sigma(+f)$·   normal $= N$
              $+\sigma(+f)$ and $+2\sigma(+2f)$:   above normal $= A$
              $+2\sigma(+2f)$ and $+3\sigma(+3f)$: greatly above normal $= GA$
above: $+3\sigma(+3f)$ extremely above normal $= EA$

At State College, Pennsylvania,[6] 53 years' observations of the mean daily minimum of the coldest month were available. The probable error is: $f = \pm 2.9$ F°.

Hence the following classification (as indicated in Table 7). The class limits are derived from the 53 years 1886 to 1938. It should be kept in mind that these limits are changed if another period is used On the other hand, the variability of the limit is smaller by far than that of the values of the element itself. In climatographies, such classifications of temperature, precipitation, etc., should be given for each place with about 50 or more years' observations.

From Table 7, one concludes that a deviation from the mean daily minimum of > -8.7°F is to be considered a rare event, with a theoretical probability of two occurrences in 100 years. Consequences from such low temperatures (ruin of winter wheat, of rosebushes, etc.) could be judged as caused by a "superior force."

TABLE 7. Classification of Deviations from the Mean Daily Minimum Temperature at State College, Pa , Based on the Probable Error

(1) Charac- teristics	(2) Classification	(3) Observed frequency	(4) Calculated frequency	(5) Obs − Calc *
		%	%	%
ES	< -87 F°	1 88	1 80	+0 08
GS	-87 to -58	7 55	6 72	+0 83
S	-58 to -29	18 87	16.13	+2 74
N	-29 to $+29$	45 28	50 00	−4 72
A	$+29$ to $+58$	18 87	16.13	+2.74
GA	$+58$ to $+87$	7.55	6.72	+0 83
EA	$> +87$	0 00	1 80	−1.80

* Column 3 minus column 4

[6] V. Conrad, "Investigation into Periodicity of the Annual Range of Air Temperature at State College, Pennsylvania," *The Pennsylvania State College Studies*, No. 8 (State College, 1940), pp 10 ff. See also F H Chapman, *Meteorological Office, London· Professional Notes*, 1919, no. 5; and V Conrad, *Meteorologische Zeitschrift*, 1921, p. 91.

The example in Table 7 shows that the "probable error" has the advantage that the intervals are rather smaller than at the standard deviation. This is valuable for climatical purposes since limits of rare cases are better adjusted to common feeling.

The theoretical frequency distribution is represented as a full line in Figure 6, and the crosses mark the observed frequencies at the midpoints of the class intervals. It should be pointed out that

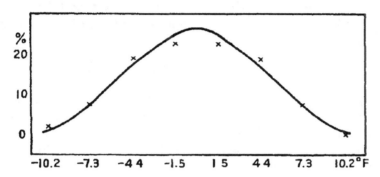

Fig. 6   Frequency distribution (per cent) of average daily minimum temperatures in State College, Pa

Calculated: Full line.   Observed. x.*   (After V. Conrad)

* The numbers below the abscissa mean the midpoints of the class intervals. The normal class is divided into two class intervals.

the observed frequency curve fits the normal distribution very well. The small departure within the modal class interval is called *positive excess.*

III. 4. Variance.   Coefficient of Variation

Before continuing the discussion of how far climatological series show a more or less normal frequency distribution, one may mention two further measures of the *dispersion* (scatter) of a variate. Both, e.g., are used in the papers and reports of the agricultural climatological department of British India.[7] They are directly derived from the standard deviation (σ).

The first is the variance (Va):

$$Va = \frac{\Sigma d^2}{n}$$

[7] See, for instance, P C Mahalanobis, *Report on Rainfall and Floods in North Bengal* (Calcutta, 1927).

It is the square of the standard deviation, and is somewhat easier to calculate. The indications of the variance react a little more strongly to small changes of the distribution of deviations, as exemplified in the left side of Table 6 (Vancouver):

$$Va = \frac{\Sigma d^2}{n} = \frac{712}{10} = 71.2$$

The other measure is the *coefficient of variation* (*CV*), that is, the standard deviation expressed as a percentage of the arithmetic mean ($\dot{m}$)·

$$CV = 100\,\frac{\sigma}{\dot{m}} = 100\,\frac{8.44}{108} = 7.81\%$$

if the numerical values are taken from Table 6 (Vancouver).

The "coefficient of variation" should be more frequently used in climatological statistics; it permits an exact comparison of the values characterizing the dispersion because it is reduced to an equal arithmetical mean.

III.   5. Summary of Statistical Concepts

Thus far the following statistical expedients have been explained:

Frequency Distribution		Mean Deviation	(μ)
Median		Standard Deviation	(σ)
Quartiles		Normal Distribution	
Deciles [8]		Probable Error	(f)
Arithmetic Mean	($\dot{m}$)	Variance	(Va)
Deviations	(d)	Coefficient of Variation	(CV)

III.   6. Relation between μ, σ, and f. Cornu's Theorem. Unilaterally and Bilaterally Limited Variates

In the foregoing discussion only the relation of standard deviation (σ) and probable error (f) was mentioned:

$$f = 0.6745\sigma$$

Standard deviation (σ) and mean deviation (μ) are connected by

---

[8] To my knowledge, *sextiles* (6 parts) and *percentiles* (100 parts of a sample) are not usual in climatological investigations For the rest, the method of calculation is identical to that used for medians, quartiles, etc.

the following simple equation:

$$\sigma = \sqrt{\frac{\pi}{2}}\,\mu = 1.253\mu$$

As a rough approximation one can use

$$\sigma = 5/4\mu$$

These relations are exactly valid only in the case of a normal distribution, and approximately valid if the departures from the normal distribution are not too great. If the size of the climatological sample is not too small (lower limit perhaps 25 items), σ can be computed with the above equation if there is some likelihood that the variate can be represented more or less by a normal frequency distribution: An example may be taken from Table 6. The mean deviation μ (column 3) is:

$$\mu = \pm\, 7.2$$

therefore

$$\sigma = 1.253\mu = \pm\, 9.02$$

instead of 8.44. This means an error of 7% of the true value. The agreement mentioned is not too good, because the number of items is much too small. The equation

$$\sigma = \sqrt{\frac{\pi}{2}} \cdot \mu$$

can be written

$$\frac{2\sigma^2}{\mu^2} = \pi = 3.1416$$

This expression is called Cornu's Theorem and has often been used as a solution of the problem, whether or not one may assume that the climatological elements can be represented by a normal distribution. This inversion of the problem may be illustrated by the following examples. The deviations of

125 mean winter temperatures at Vienna, Austria (Central Europe), yield

$$\frac{2\sigma^2}{\mu^2} = 3.11$$

125 mean summer temperatures at Vienna, Austria, yield

$$\frac{2\sigma^2}{\mu^2} = 3.25$$

130 average annual temperatures at Paris, France (Western Europe), yield

$$\frac{2\sigma^2}{\mu^2} = 3.16$$

110 average annual temperatures at Milan, Italy, yield

$$\frac{2\sigma^2}{\mu^2} = 3.11$$

130 average January air pressures at Paris yield

$$\frac{2\sigma^2}{\mu^2} = 3.13$$

From the ten years' data of the last killing frost in Vancouver (Table 6) it follows that:

$$\frac{2\sigma^2}{\mu^2} = \frac{2 \times 8.44^2}{7.2^2} = 2.75$$

The differences between the other results are only 4% of π.

One might think that in general the series of deviations of a climatological element shows the normal distribution. An example of this behavior is given in Table 7 and Figure 6. On the other hand, it must be emphasized that the fulfilling of Cornu's theorem is *only a necessary, but not a sufficient condition* of normal distribution. This means that if Cornu's criterion is not satisfied, no normal distribution can be expected; but the criterion can be fulfilled even if the deviations obey another law of distribution. Therefore the inferences from Cornu's theorem are not binding.

Finally, one point, often overlooked in textbooks and in scientific papers, must be stressed. A normal distribution is possible only if there is—as stated earlier—no effective reason that the deviations should not take all values from $-\infty$ to $+\infty$. This requirement is more or less fulfilled in the case of pressure, of air temperature, and of some derived elements.

Many elements have zero as the limit on one side; e.g. precipitation, wind velocity, etc. In spite of this fact a more or less normal frequency distribution and inferences from it are not excluded if the frequency curve shows only relatively small frequencies in the vicinity of the given limit (e.g., zero).[9]

[9] Naturally also skewness and kurtosis (peakedness, excess) have to be considered besides dispersion if an exact statement is to be made as to whether or not there is a Gaussian distribution. See the rain example in R A Fisher, *Statistical Methods*, p. 54 ff.

The precipitation during March in Helwan, Egypt (Table 8), offers an example of an extreme case.   The deviations yield:

$$\sigma = \pm 8.03 \text{ mm}$$

The arithmetic mean, however, is 6.3 mm of rain in March on the average for 21 years.

TABLE 8   RAIN IN MARCH AT HELWAN, EGYPT
(29.9°N, 31 3°E, 38 FT.)

Year	A Amounts (mm)	B Deviations	Squares of the deviations*
1904	0	−6 3	39.7
5	1 9	−4 4	19.4
6	4 8	−1 5	2 2
7	9 1	2 8	7 8
8	24.8	18.5	342 2
9	0	−6.3	39.7
10	7.8	1.5	2.2
11	0 4	−5 9	34 8
12	0	−6 3	39 7
13	1 6	−4 7	22.1
14	0	−6 3	39.7
15	0	−6.3	39.7
16	25 8	19 5	380 2
17	0	−6 3	39 7
18	13.0	6.7	44.9
19	0	−6 3	39.7
20	7 2	0 9	0 8
21	9 1	2 8	7 8
22	7 5	1 2	1 4
23	0	−6 3	39 7
24	19 4	13 1	171 6
Σ..		133 9	1355 0
m̃	. 6 3		

C   Deviations Grouped into Class Intervals

Class interval	Midpoint, mm	Frequency %
−7.0 to −1.1 mm	−4 0	57
−1 0 to 4 9	2 0	24
5 0 to 10.9	8 0	5
11 0 to 16 9	14 0	5
17 0 to 22.9	20 0	9

* Rounded off to the first decimal place.

The negative deviations are crowded within the first class interval (Table 8, C) and the frequency curve has nothing to do with the normal distribution as seen in Figure 7.   These conditions are exaggerated even more if the variate is bilaterally limited.

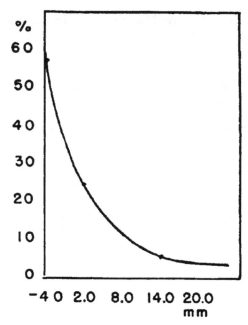

FIG 7 Frequency curve of rainfall in March (21 years) at Helwan, Egypt On the X-axis Midpoints of the class intervals of the deviations (mm)

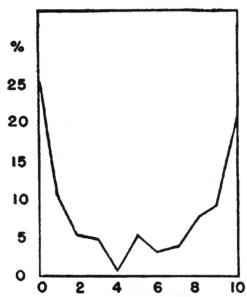

FIG 8 Frequency distribution (per cent) of the degrees of cloudiness in ten September months at 9 p m , at Prague, Czechoslovakia

Perhaps the best example is the element *cloudiness* with its two limits: zero for cloudless sky, and ten for overcast sky. Figure 8 shows the frequency distribution of cloudiness (in per cent) for ten months of September in Prague (50.1°N, 14.4°E, 663 ft), at 9 P.M. The extremes are by far the most frequent values in contradistinction to the normal distribution, because of the bilateral limitation of the variate and of regional climatic properties.[10] The arithmetic mean lies at about cloudiness 5. The arithmetic mean here loses its regular sense and indicates more or less the degree of mixing of the extreme values.

To go back for a moment to the example of the March rain in Helwan: Cornu's theorem can be proved by means of the data given in Table 8. We get:

$$\frac{2\sigma^2}{\mu^2} = 3.17$$

or a value only 1% greater than Ludolf's number. This is a typical example of a series of an element with a frequency distribution different on principle from the normal distribution. Nevertheless, Cornu's theorem is nearly fulfilled. We repeat that it is a necessary and not a sufficient condition.

III. 7. Higher Characteristics (Skewness)

The higher characteristics are used to analyze frequency distributions which are not identical with the normal distribution. For climatological purposes the asymmetry of the frequency curves is of interest. The term for it is *skewness*. In contradistinction to the symmetrical normal distribution, here arithmetic mean and mode no longer coincide. The arithmetic mean is much affected by extreme values of the variate; the mode is not affected at all by the extremes. Therefore mean and mode cannot coincide. The greater the distance is between these two characteristics, the greater is the asymmetry, the skewness (*Sk*). According to what has been said before, the mean moves toward that side which shows the greater and more frequent extreme values. Therefore a frequency curve with the mean at the right side of the mode is called *positively skewed* (showing *right skewness*), while in the opposite case, it is *negatively skewed* (showing *left skewness*). Right and left skewness may be seen in Figures 9a and 9b.

[10] The statistics are taken from L W. Pollak, in *Prager Geophysikalische Studien*, vol. VI (Prague, 1931), p 11.

K. Pearson gives as a measure of skewness the difference between mean and mode.   This difference has to be reduced to equal dispersion.   Thus the equation results:

$$Sk = \frac{\text{Mean} - \text{Mode}}{\sigma}$$

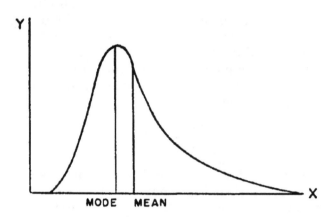

FIG. 9a.   A right (positively) skewed frequency distribution showing the theoretical position of mode and mean.   (After Arkin and Colton)

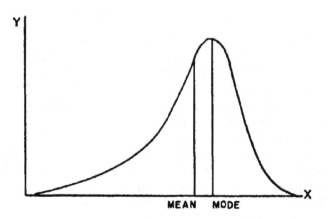

FIG. 9b   The same as above, but left (negatively) skewed

It is useful to know, for instance, how far temperature deviations of one sign predominate as to quantity and to frequency. The trouble is that the position of the mode is rarely calculated, because of the great labor involved for long series.

Less burdensome is Köppen's method of calculating the asymmetry (A).   The equation reads:

$$A = 1 - \frac{2n_b}{n}$$

where n_b means the number of items below the mean, and n the total number of items.   For the case of symmetry, $A = O$ since $n_b = n/2$.

Example: C. F. Brooks gives the annual amounts of precipitation [11] for $n = 80$ years, at Philadelphia, Pa., expressed in deviations.   The mean is added.

A simple counting yields in a few minutes

$$n_b = 43$$

therefore

$$A = 1 - \frac{86}{80} = -0.075$$

Compared with other results, the negative asymmetry is rather high.   Negative deviations are a little more frequent.   The expectation of a relatively dry year is somewhat greater than that of a wet year.

The great advantage of Köppen's method is that it can be computed easily, and the A from different places are comparable with one another.   On the other hand, it should not be forgotten that Köppen's method considers only the frequency of the signs, and that the amounts of the deviations are ignored.   This is the great disadvantage of this method compared with the skewness method.

Other characteristics of the frequency distribution, such as the peakness of the curve of frequency distribution, moments, etc., are of less interest to climatologists, and are not discussed here. [12]

III.   8. DIFFERENT KINDS OF VARIABILITY USED IN CLIMATOLOGY

1) Mention was made earlier of average variability, which gives a good measure of the variations to which the climatological element is subject.

[11] R. De Courcy Ward and C. F Brooks, "The Climates of North America," in Köppen-Geiger *Handbuch der Klimatologie*, vol II, pt. J, p. 283.

[12] See R. A Fisher, *Statistical Methods*, H. Arkin and R. R. Colton, *An Outline of Statistical Methods* (4th ed , New York, 1939); H L Rietz, *Handbook of Mathematical Statistics* (New York, 1924).

The average temperature of January has an average variability $\Sigma|d_i|/n$:

±0.7 F° in Batavia, Java (6.2°S)
±9.7 F° in Green Harbor, Spitsbergen (78.0°N)

The average variability of the mean temperature of the months, is indeed climatically very characteristic. In the Arctic climate, the average variability of the temperature in January is about 14 times greater than that in the inner tropic belt.

2) If the series of observations is rather short (<10 years) the absolute range of variation of the series gives a first approximation of the variability of the element.

3) The average variability is independent of the sequence of items. This is, however, of special climatological interest. It is important in judging of a climate, for instance, whether the changes from one day to another are rather smooth or very abrupt.

This method is not restricted to equal sections of time (e.g., the day). It can be applied also to equidistant places along a straight line, running, for instance, from the coast inland. Special variabilities, regarding the sequence of the places in a given direction, could yield new characteristics of a special contrast in the climate of a region.

Thus a measure of variability which considers the quantity as well as the sequence of the items of the series of numbers characterizing the variations of a climatic element is necessary.

Table 9 offers an example of the "average variability" (AV) computed from the series in part A: $AV = \pm 5.0$.

In part B another variability (SV) is calculated in the following way:

$$SV = [|a_1 - a_2| + |a_2 - a_3| + \cdots + |a_{n-1} - a_n|] : (n - 1)$$

if n equals the number of items. This variability may be called *inter-sequential variability*, because its value is determined by the quantity of the items as well as by their sequence. In the example, $SV = \pm 8.8$.

For the sake of clarity, the series is arranged as an array in part C of Table 9, to show how great the influence of the order of numbers is upon the quantity of SV. This is now a quarter of that in part B.

This important method of calculating a variability which takes into account the sequence of the items in the climatological series

TABLE 9. COMPARISON BETWEEN AVERAGE VARIABILITY (AV) AND
"INTER-SEQUENTIAL" VARIABILITY (SV)

n = number of items d = deviations from arithmetic mean ($\bar{m}$).
D = Difference between two consecutive items

A

Original order	a_1	a_2	a_3	a_4	a_5	a_6	a_7	a_8	a_9	a_{10}
of 10 numbers	9,	8,	17,	2,	9,	20,	0,	7,	12,	16; $\bar{m} = 10$
deviations	-1,	-2,	$+7$,	-8,	-1,	$+10$,	-10,	-3,	$+2$,	$+6$

$$\Sigma(d_+) = 25, \qquad \Sigma(d_-) = 25$$

$$AV = \pm \frac{\Sigma|d|}{n} = \pm 5.0$$

B

Computation of SV	1,	-9,	15,	-7,	-11,	20,	-7,	-5,	-4
from the series A									

$$\Sigma(D_+) = 36, \qquad \Sigma(D_-) = -43$$

$$SV = \frac{\Sigma|D|}{n-1} = \frac{+36+43}{9} = \pm 8.8$$

C

Series of A arranged as array	0,	2,	7,	8,	9,	9,	12,	16,	17,	20
Differences between consecutive items	-2,	-5,	-1,	-1,	0,	-3,	-4,	-1,	-3	

$$\Sigma(D_+) = 0, \qquad \Sigma(D_-) = -20$$

$$(SV)' = \frac{\Sigma|D|}{n-1} = \frac{+0+20}{9} = \pm 2.2$$

can be applied to each element as far as it represents a scalar quantity completely specified by one number, in contradistinction to directional quantities, which have direction as well as magnitude (wind, motion of clouds, motion of the water in an ocean current, etc.).

The best-known example of an inter-sequential variability is the interdiurnal variability. This is calculated by two methods:

1) Differences of the consecutive daily means of the element;
2) Differences of the element at a fixed hour of consecutive days.

The advantage of the first method is that, for instance, the mean daily temperature is derived from at least two observations (e.g., the extremes), so that incidental errors are not so likely. On the other hand, average values can always conceal interesting phenomena.

The disadvantage of the second method is that, for instance, the interdiurnal variability of temperature has a well-accentuated daily variation. The greatest trouble is that the time of the extremes is variable in the course of the year.

In Central Europe (Potsdam), the maximum interdiurnal variability occurs at 6 A.M. in October, and at 4 P.M. in April. This daily variation reaches its maximum in July, when the ratio maximum/minimum is 1.8. Much caution should be exercised before drawing conclusions from investigations in interdiurnal variability taken from a fixed hour.

Tables 10 gives a nearly complete example of an investigation in interdiurnal variability of temperature.[13]

TABLE 10. DIFFERENT STATISTICS REGARDING INTERDIURNAL VARIABILITY OF TEMPERATURE

Col 2 = Interdiurnal Variability of Temperature (5 A M) (IDV) at Mount Washington, N H. (6270 feet)
Col 3 = Average Coolings (C)
Col 4 = Average Warmings (W)
Cols 5 and 6 = Average Numbers of Consecutive Days with Increasing (WD) and Decreasing (CD) Temperature
Col 7 = Length of Temperature Surges in Days (S)

(After V Conrad)

(1) Month	(2) IDV F°	(3) C F°	(4) W l·°	(5) CD days	(6) WD days	(7) S days
Dec	8 7	9 2	8.3	1.84	1 92	3 76
Jan	11 4	13 8	9 6	1.44	1 86	3 30
Feb	10 1	9 9	10 4	2 36	2 45	4 81
Mar	9 8	10 4	9 2	1 79	1 92	3.71
Apr	6 2	6 6	5.9	1 62	2 13	3 75
May	5 9	6 3	5 6	1 79	2 13	3 92
June	5 1	5 2	5 0	1 79	1 95	3 74
July	4 0	4 4	3 7	1 74	2 13	3.87
Aug	4 2	4 4	4 1	2.05	2 40	4 45
Sept	6 3	7 5	5 2	1 87	2 13	4 00
Oct.	8 0	8 3	7 7	2 09	2 23	4 32
Nov.	8 0	8 5	7 5	1.96	2 04	4 00
(m) . .	7 3	7.9	6 8	1 86	2 11	3.97

Column 2 contains the interdiurnal variability (IDV) at 5 A.M. The values are averages of 3 years. Irregularities of the annual course of IDV, as well as of the elements in the other columns, are to be explained by the shortness of the period under discussion.

[13] V. Conrad, "The Interdiurnal Variability of Temperature on Mount Washington," *Transactions of the American Geophysical Union* (Washington, 1942), pt. II, p 279.

The statistics which lead to the IDV (which ignores the signs) enable one to calculate separately the average amount of temperature—increasing (W) (col. 4), or decreasing (C) (col. 3)—from day to day. The increase of temperature from one day to the next might be termed warming; the decrease, cooling.

Because in the problem in question the amounts of cooling and of warming are not effectively influenced by radiation processes, they are a measure, or at least an indication, of advected cold and warm air masses. Therefore, these statistics are valuable for climatology as well as for long-range forecasting, etc. When the IDV is calculated, average and maximum values of cooling and warming should be published. This can be done with little or no supplementary labor.

A further valuable addition is represented by computing the number of consecutive days with increasing temperature (WD) (column 6), and with decreasing temperature (CD) (column 5). The individual numbers are averaged for each month.

Column 7 shows the sums of $CD + WD$: that is, the average time in days for an average, consecutive increase and decrease. This seesaw can be called a *temperature surge*, and column 7 (S) gives the average duration of the surges in days.

Once more, it should be emphasized that the inter-sequential variability can and should be used when different climatological elements are discussed. For lack of space, only one example of temperature and the time-unit of one day is given in this chapter.

For one purpose or another, an *interhourly variability* may be useful. As an example, the wind velocity (regardless of the direction) could be studied from this standpoint, if the establishment of windmills in a given locality were planned.[14] B. M. Varney [15] uses an *interannual variability* to characterize the rain conditions of California. Annual and seasonal amounts are both handled by this method. It goes without saying that monthly average data for air pressure, temperature, cloudiness, vapor pressure, precipitation, depth of snow, etc., can be studied also by interannual, inter-seasonal, etc., variability.

[14] See also V. Conrad, "Interhourly Variability of Temperature at Mount Washington, N. H.," *Transactions of the American Geophysical Union*, 1943, p. 122.

[15] B M Varney, "Seasonal Precipitation in California and its Variability," *Monthly Weather Review*, 1925, pp. 148–163, and 208–218.

III. 9. Absolute and Relative Variability and Other
Measures of Variations

If amounts of such as precipitation are considered, the average
variability $\dfrac{\Sigma|p_i - \bar{p}|}{n} = AV$ depends naturally upon the arithmetic
mean $(\bar{p})$.[16]

The normal annual precipitation in Cairo, Egypt, is 1.34 in.;
the variability is 0.67 in. This is large, because the annual precipitation varies between one half and three halves of the mean.
The same variation in Rangoon, Burma (17°N, 96°E), with a
"normal" precipitation of about 100 in., would be negligible indeed.

Therefore, a new concept is necessary, no longer dependent on
the arithmetic mean: the *relative variability*. This is the absolute
variability expressed in per cent of the arithmetic mean.

$$V_r = 100\,\frac{AV}{\bar{p}} = \frac{100}{n}\frac{\Sigma|p_i - \bar{p}|}{\bar{p}}$$

The relative variability V_r would reduce the absolute variability
AV to the unit of annual precipitation only if the correlation between the two variants, V_r and $\bar{p}$, were linear. This assumption
does not hold, as is shown from the observations.

Statistics from about 360 stations scattered over the earth
yielded a correlation between V_r and normal rainfall. This is
represented in Figure 10 by a hyperbolic curve. Therefore, the
influence of the normal sum of precipitation becomes weak only
beyond the turning-point of the curve, i.e., if the normal rainfall is
larger than about 28 inches. Then, in a first rough approximation,
V_r is practically no longer dependent upon $\bar{p}$. It should be stressed
that about 40% of the earth's surface has less than 28 inches precipitation, and 30% has less than 20 inches. Conclusions drawn
from comparing values of V_r for different places in these vast
regions of small annual precipitation are inaccurate and misleading.[17] A way out of these difficulties is offered by the representa-

[16] If the arithmetic mean is calculated from a longer period, perhaps at least 25
years, it is called *normal value* or *normal* The Meteorological Office in London published a *Book of Normals* based on a period of 35 years (see XII 3 f)

[17] For more details, see V. Conrad, "The Variability of Precipitation," MWR,
vol. 69 (1941), pp. 5–11. Statistics taken from E. Biel, "Die Veränderlichkeit der
Jahressumme des Niederschlags auf der Erde," *Geographisches Jahrbuch aus Oesterreich*,
14/15, 1929.

tion of anomalies, treated later (XIV, 7). In regions with an annual precipitation greater than about 20 to 28 inches, the values of V_r can be compared with one another without serious error.

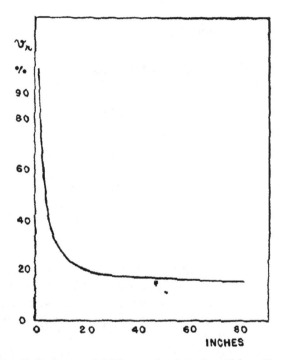

FIG. 10.   Relative variability and rainfall   (After V Conrad)*

* v_r in the diagram should read V_r.

Since the calculation of V_r is rather troublesome for long series of observations in many investigations of precipitation conditions, a simpler measure of the variability is commonly used. G. Hellmann introduced an expression which is called *ratio of variation* (Q).[18] It is nothing but the quotient

$$Q = \frac{M}{m}$$

where $M =$ the largest annual amount in the series and $m =$ the smallest. This measure should *not* be used, since Q becomes *infinite* in many places located in the desert belts.

[18] G. Hellmann, *Veroffentlichungen Preuss. Met. Inst.* III (1909), no. 1.

E. Gherzi used another measure of variation: [19]

$$Q = \frac{M - m}{\bar{p}}$$

It is a great improvement over the measure M/m.[20]

III.   10. Some Applications of the Method of Random Samples

III.   10. a. *Precipitation*

Though the following inferences are restricted to one or another element (precipitation and fog) similar methods could be applied to obtain results for other elements.

At some thousands of climatological stations, eye observations are made from one to three times a day, but no continuous records are kept.   One assumes that the observer enters the international sign for rain or snow in his register when it rains or snows at the precise time of the observation.   Then, the observation can be valid for this time as a random sample of the actual weather. Starting from this idea, W. Koppen derived from these instantaneous observations data which, otherwise, could be taken from continuous records only.   Probability has previously been defined as the ratio of the number of favorable or desired events to the total number of possible events.   If one is interested in precipitation, observations, just at the moment of rain or snow, are the favorable events.   In a given period (e.g., a month), n observations are made.   Among these, r observations occur exactly during the precipitation,   Then, the proper fraction $r/n = p$, is the *absolute probability of precipitation,* "absolute," because independent of the selected unit of time.   The absolute probability of precipitation is the basis of Koppen's estimates.

The following statement and some simple inferences give a survey of the quantities in question and a demonstration of how to use them.

n = Number of all observations within the chosen period (e.g., month, season)

r = Number of all observations with precipitation

---

[19] $\bar{p}$ = average amount of precipitation of the period in question.
[20] E. Gherzi, *Étude sur la pluie en Chine* (Observatoire Zi-Ka-Wei, 1928).

N = Total number of hours in the chosen period (e.g., 720 hours in month of June)

d = Number of days with precipitation within the chosen period

h = Amount of precipitation in the period in question

$p = \dfrac{r}{n}$ = Absolute probability of precipitation

Then:

$D = pN$ = Probable total duration of precipitation in hours

$\dfrac{D}{d} = \dfrac{pN}{d}$ = Average duration of precipitation, per day with precipitation, in hours

$\dfrac{h}{d}$ = "intensity" of rain (average amount of rain per rainy day)

$\dfrac{h}{D} = h \cdot \dfrac{d}{pN}$ = Average amount of rain in one rainy hour

Example: At the observatory in Batavia (Java), not only were all the necessary observations made by eye and by self-recording gauges, but also the data were published in full detail, so that the method described can be tested to see how far it is in agreement with observation.

January 1915 [21] was the month of the test (dry period).
Regular observations were made three times a day: $n = 93$
By counting, it was determined that. $r = 10$
Therefore: $p = r/n$ $= 10/93 = 0.1075$
Further: $N = 31 \times 24$ $= 744$ hours
Hence: $D = pN$ $= 80$ hours with rain

Observed: $d = 29$
Therefore: $D/d = 80/29$ $= 2.76$ hours
Observed: $h = 13.07$ inches
Hence: $h/d = 13.07/29 = 0.45$ inches/rainy day
and $h/D = 13.07/80 = 0.163$ inches/rainy hour

From Table 11 it can be seen how well this method given by Koppen agrees with observation. The computation is simple and the method is easy to understand. On the other hand, it supplies information about data which—as stated before—can usually be

[21] *Yearbook of Batavia, Java.*

TABLE 11. METHOD OF RANDOM SAMPLES APPLIED TO RAINFALL AT
BATAVIA, JAVA

Number of rainy hours		Average duration of rain in one rainy day		Average amount of rain in one rainy hour	
D obs	D calc.	D/d obs hours	D/d calc. hours	h/D obs. in	h/D calc. in.
75 3	80	2.6	2 8	0 173	0.163

obtained only from continuous records. These ideas could be used,
with the necessary modifications, for other climatological elements.

III. 10. b. *Cloudiness (C) and Duration of Sunshine (S)*

Another problem using random samples may be described.
At every climatological station, cloudiness is observed. At a few
of them, bright sunshine is recorded continuously. It goes without
saying that the average monthly duration of sunshine is most
interesting information for many purposes.

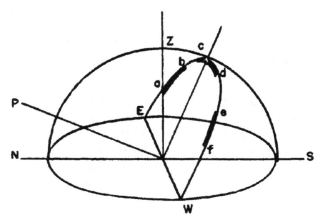

FIG. 11 The relation between cloudiness and duration of sunshine
E abcdef W = path of the sun. The parts of the path *ab, cd, ef,* are covered
by clouds

The question of how to calculate sunshine duration from average cloudiness is approximately solved by means of random samples. We assume that the path of the sun (Fig. 11, *E a b c d e f W*)
for a given day of the year is obscured by clouds for the portions
$\overline{ab}$, $\overline{cd}$, $\overline{ef}$. If these parts are known from the records of a sunshine
recorder, a simple calculation gives the percentage, *C*, of the total

length of the path of the sun which is covered by clouds. It is clear that the rest of the path $S = 100 - C$.

Now, if the assumption is made that the conditions along the path of the sun are a random sample, valid for the entire visible sky, the problem is solved in principle. In reality, this hypothesis cannot be made for an individual day, only for an average day within a period of at least one month. Even then, the assumption remains a rough approximation, as will be grasped from the examples in Table 12. They show that the annual and monthly values of S and C yield sums which indeed generally differ less than 10% from 100. This agreement is not bad.

The simple formula, $S + C = 100\%$, has been improved by theoretical considerations. The resulting formula is not tested

TABLE 12   THE EQUATION $S + C = 100\%$ COMPARED WITH OBSERVATION RESULTS IN CALIFORNIA AND IN THE MEDITERRANEAN REGION

(S in % of possible duration; C in % of the visible sky)

		Jan	July	Year
		%	%	%
San Francisco, Calif	S	53	69	65
37.8N, 122.7W	C	54	37	42
	$S + C$	107	106	107
San Diego, Calif	S:	67	68	68
32 7N, 117 2W	C:	40	34	36
	$S + C$	107	102	104
Sacramento, Calif	S.	46	96	74
38 6N, 121.4W	C	59	08	31
	$S + C$:	105	104	105
Messina, Italy	S:	37	75	52
38 2N, 15 6E	C·	61	28	46
	$S + C$.	98	103	98
Athens, Greece	S:	49	81	60
38 0N, 23 7E	C·	55	11	40
	$S + C$:	104	92	100
Helwan, Egypt	S.	70	90	82
29.9N, 31.1E	C:	41	06	23
	$S + C$·	111	96	105

for different climates, and its evaluation is relatively complicated, so that, generally, the climatographer would prefer not to use it.[22]

Finally, in connection with this problem, another definition may be mentioned:

$$B = 100 - C$$

where C is again the cloudiness in per cent of the visible sky; B can be called the *brightness of the sky*, and is used sometimes in meteorological and climatological papers.

[22] See V. Conrad, "Die klimatologischen Elemente und ihre Abhangigkeit von terrestrischen Einflüssen," *Handbuch der Klimatologie*, vol I (Berlin, 1936), p. 450, and A. Wagner, "Beziehungen zwischen Sonnenschein und Bewölkung in Wien," *Meteorologische Zeitschrift*, 1937, pp. 161–167.

CHAPTER IV

SOME PROBLEMS OF CURVE FITTING AND SMOOTHING OF NUMERICAL SERIES

IV. 1. THE STRAIGHT LINE

IN CLIMATOLOGY, one is often confronted with the problem of representing an element as a function of an independent variable, such as altitude, latitude, distance, etc. Each value of the variable element is related to a certain value of the independent variant. The first step is to arrange the values of the element according to increasing values of the independent variable. If there is a greater number of values, it may be advantageous to divide the data into groups, and to average either variant within each interval.

The variables are plotted on the X- and Y-axis, respectively, and a line is drawn freehand through the resulting points.

This procedure may suffice for survey of the correlation of both variants. In any case, the type of curve can be recognized; in climatology, one may often assume it to be a straight line.

TABLE 13. THE RELATIONSHIP BETWEEN THE AMPLITUDE OF THE ANNUAL VARIATION OF AIR PRESSURE (a_1, in mm mercury) AND THE ALTITUDE ABOVE SEA LEVEL (h in hectometers, 1 hectometer = 100 m)

No.	(1) h (hm)	(2) a_1 (mm Hg observed)	(3) a_1 (calculated by the semi-average method $a_1 = 0\ 29 + 0\ 16h$)	(4) Obs −calc (2-3)	(5) a_1 (calc with $a_1 = 0\ 36 + 0\ 16h$)	(6) Obs −calc
1	0 5	0 28	0 37	− .09	0 44	− 16
2	2 7	1.08	0 72	+ 36	0 79	+ 29
3	4 6	1.04	1.02	+ 02	1 09	−.05
4	6 1	1 24	1 26	−.02	1.33	−.09
5	7 8	1 38	1 54	− 16	1.61	− 23
6	9.0	1 72	1.73	−.01	1 80	− 08
7	16 4	3 04	2 91	+ 13	2.98	+.06
8	23 4	4 26	4 03	+ 23	4 10	+.16
9	30.5	5 34	5.17	+ 17	5 24	+ 10
$\Sigma(+)$				+91		+61
$\Sigma(-)$				−28		−61
				+63		0

An example illustrates this problem· Records of the whole-year amplitudes of air pressure at 31 stations are correlated with the altitude of the respective places. The amplitudes are arranged in 9 groups according to increasing heights. The averages (hectometers) of the intervals are indicated in column 1 of Table 13. The averages of the respective amplitudes are contained in column 2.

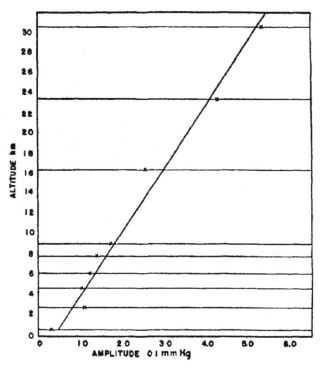

Fig. 12.   Annual range of air pressure (abscissa) and altitude (ordinate). (After V. Conrad)

Starting from these two series of numbers, one derives the points marked on the graph (Fig. 12). A straight line fits the observations best.

The analytical equation of a straight line has the form:

$$a_1 = A + Bh$$

where a_1 means the amplitude of the whole annual variation, h the altitude, and A, B, constants which have to be evaluated. For this reason 9 equations are available from Table 13.

$$\text{No. 1} \quad 0.28 = A + 0.5\,B$$
$$\text{No. 2} \quad 1.08 = A + 2.7\,B$$
$$. \qquad . \qquad .$$
$$\text{No. 9} \quad 5.34 = A + 30.5\,B$$

Here is just the case to use the method of least squares.

As was said above, however, as a rule it does not pay to apply this rather wearisome method to climatological problems. Usually, the accuracy of the observations is not sufficient to use least squares.[1] Therefore a method of quick approximation is given

[1] In the case of a linear relation, the method of least squares yields relatively simple equations for the constants A and B of the equation $a = A + Bh$, the computation of which is not too laborious.

$$A = \frac{\Sigma h \cdot \Sigma(ah) - \Sigma a \cdot \Sigma h^2}{(\Sigma h)^2 - n\Sigma h^2} \qquad B = \frac{\Sigma h\,\Sigma a - n\Sigma(ah)}{(\Sigma h)^2 - n\Sigma h^2}$$

These formulas, applied to the example of Table 13, yield (1) for the calculation of the numerator of A:

$$\Sigma h = 101.0; \qquad \Sigma(ah) = 354.158;$$
$$\Sigma h \cdot \Sigma(ah) = 35769\ 958,$$
$$\Sigma a = 19\ 38, \qquad \Sigma h^2 = 1954.52$$
$$\Sigma a \cdot \Sigma h^2 = 37878\ 5976$$

(2) for the calculation of the numerator of B

$$\Sigma h\ \Sigma a = 1957\ 38; \qquad n\ \Sigma(ah) = 3187\ 422$$

(3) for the denominator of A and B.

$$(\Sigma h)^2 = 10201\ 0, \qquad n\ \Sigma h^2 = 17590\ 68$$

and finally:

$$A = +\ 0\ 2853$$
$$B = +\ 0\ 1664$$

so that the equation, got by the method of least squares, reads:

$$a_1 = 0\ 285 + 0\ 166h \qquad\qquad\qquad\text{(I)}$$

The semi-average method, given in the text, resulted in

$$a_1 = 0\ 29 + 0\ 16h \qquad\qquad\qquad\text{(II)}$$

The values of A are nearly identical, the values of B differ for about 4%. For further correction, the straight line defined by equation (II) is shifted parallel to itself increasing A from 0.29 to 0.36 and the equation reads

$$a_1 = 0.36 + 0.16h \qquad\qquad\qquad\text{(III)}$$

This step is advantageous in so far as it makes the sum of the positive deviations equal to that of the negative ones by a minimized arithmetic. In the present example it would have been more nearly correct to leave the constant A unchanged and to diminish the angle between the straight line and the abscissa as it can be seen from the exact equation (I). On the whole, the agreement between the approximations given by the semi-average method and similar methods, on the one hand, and the exact result of the method of least squares, on the other hand, is generally sufficient for climatological purposes. The reader who takes pains with making a diagram on graph paper to a scale even twice that of Figure 12 will become aware of how close the three lines run to each other.

here instead, the *semi-average method*.   The first half and the second half of the values of both variables are averaged, so that two pairs of coordinates result, which have to be introduced into the general equation of a straight line.

In the special case of Table 13, the number of items is relatively small and odd.   In order to avoid asymmetry and unequal weight, item no. 5 can be included both in the first and in the second half.

The first half then contains items nos. 1 to 5; and the second half, nos. 5 to 9.

The two equations read:

$$1.004 = A + 4.34\,B$$
$$3.148 = A + 17.42\,B$$

Therefore $A = 0.29$ and $B = 0.16$ and the straight line is given by $a_1 = 0.29 + 0.16\,h$.

The nine values of h (Table 13, nos. 1 to 9) introduced into the equation yield the calculated values of a_1 in column 3.   Column 4 contains the differences· Observed values (obs.) minus calculated values (calc.).   It is seen that (at the bottom of the table) the positive differences exceed the negative ones by 0.63.   This means, that the observed a_1 are, on the average, 0.07 mm Hg greater than the calculated.   Therefore, we have to shift the calculated straight line parallel to itself to the right side by 0.07 mm: in other words, increase the constant term 0 29 by 0.07.

The definitive equation now reads:

$$a_1 = 0.36 + 0.16\,h$$

The new values computed with this equation appear in column 5, and the differences "obs. − calc." in column 6.   Incidentally, the sum of the differences (with regard to the signs) is zero, which represents the ideal.   The process of shifting has to be continued until the sum becomes a minimum.[2]   In many cases it is not even necessary to use the semi-average method.

When the number of observations varies greatly from one class interval to the next, it is sometimes more advantageous to take the averages of only two intervals with the greatest number of

[2] The example is taken from V. Conrad, "The Influence of Altitude on the Yearly Course of Air Pressure," *Bulletin of the American Meteorological Society*, vol 20 (1939), p. 207.

observations, in such a way that the midpoints of the intervals may be as far apart as possible.[3]

IV. 2. Decay-Curves

In climatology, as in some other fields, the observation of the *decay* of any quantity—intensity, ratio, etc., with time, distance, height, or depth, has to be represented by an analytical equation.

Example: In a problem dealing with cloudiness [4] the decay of a ratio y is to be described which is infinite at the place of observation and decays to zero with increasing distance. These boundary conditions lead to the equation

$$y = Ax^{-B}$$

In terms of logarithms:

$$\log y = \log A - B \log x$$

Thus, the exponential equation is reduced to a linear equation regarding the two unknown constants $\log A$ and B. Plotting on logarithmic paper yields the values A and B in more or less rough

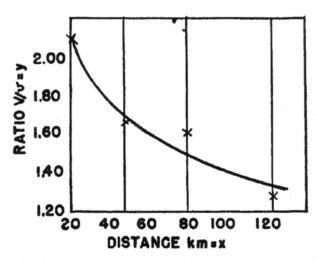

FIG. 13.   Observations represented by the formula $y = Ax^{-B}$
(After V. Conrad)

[3] For an interesting case of curve fitting, see Hans Neuberger, "Studies in Atmospheric Turbidity," *Pennsylvania State College Studies*, no. 9 (1940), p 22. There, a dot chart is presented which seemingly offers a complex problem. In reality, it is easily solved by the assumption of two straight lines instead of one curve of higher order.

[4] Taken from V. Conrad, "Zum Studium der Bewolkung," *Met. Z.S*, 1927, p. 87.

approximation. On the other hand, the problem is reduced to that of the straight line. This has been discussed above. Therefore the numerical example in Table 14 and in Figure 13 does not need further explanation.

TABLE 14   THE REPRESENTATION OF OBSERVATIONS BY AN EXPONENTIAL CURVE
$$y = Ax^{-B}$$

	(1) Distance = x (km)	(2) Ratio y obs	(3) Ratio y calc	(4) Obs − calc
	21	2 09	2 10	−.01
	48	1 65	1 69	−.05
	80	1 60	1 48	+.12
	123	1.27	1.33	−.06
Σ(+)				+ 12
Σ(−)				− 12

The solution of the logarithmic equation yields:

$$A = 4.618 \qquad B = 0.2590$$

and the equation reads:

$$y = 4.618\, x^{-0.2590}$$

IV. 3. THE EQUATION $y = B \cdot e^{Ax}$

If the dependent variable, y, has a finite value at the origin and increases with increasing x to infinite values, the equation

$$y = B \cdot e^{Ax}$$

may often represent the observations in a suitable way. Plotting on semilogarithmic paper yields the values of A and B.

In logarithmic terms the equation reads:

$$\log y = \log B + Ax \cdot \log e$$

where "log" means common (Briggsian) logarithms, $e = 2.718$, $\log e = 0.4343$. The logarithmic equation has a linear form so that A and B can be easily determined.

No general rule can be given for the choice of the right form of equation. The graphical representation remains the best guide. A collection of different forms of equations for curve fitting can be found in various books dealing with numerical calculations and mathematical statistics.[5]

IV. 4. THE QUADRATIC EQUATION

Figure 14 represents an example of a group of curves which is also important in the discussion of climatic problems. The observations averaged for 5 class intervals show the behavior of

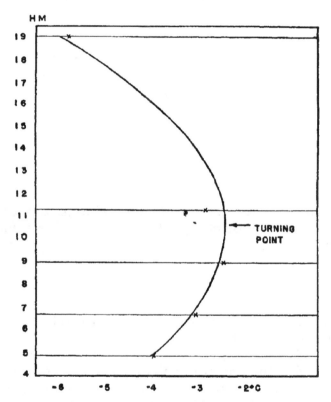

FIG. 14. Observations represented by a quadratic equation
$$(x = A + By + Cy^2)$$
(After V. Conrad)

[5] E g., in H L. Rietz, *Handbook of Mathematical Statistics* (New York, 1924), where also the transformation of the different equations to linear forms is given

Different forms of the equation for the hyperbola (e g xy = const. or $y^2 = x(m + nx)$) are often very suitable for the representation of the correlation between climatological elements and factors   Examples are given in V. Conrad, " Messung und Berechnung der Abkühlungsgrosse" (Measuring and Calculating the Cooling Power; with an English summary), *Gerlands Beitrage zur Geophysik*, vol. XXI, 1929, pp. 183–189

Another example of curve fitting by means of hyperbola equations can be found in V. Conrad, and O Kubitschek, "Die Veranderlichkeit und Mächtigkeit der Schneedecke

average winter temperature at different average heights above the bottom of an Alpine valley.[6]   The well-known phenomenon of the *inversion* is illustrated by means of the numbers in Table 15 and by Figure 14.

TABLE 15. Observations Represented by the Quadratic Equation: $x = A + By + Cy^2$ (Variation of Temperature with Height in an Alpine Valley in Winter) (h = hectometers, t in °C)

No	(1) h (hm)	(2) $t°C$ (observed)	(3) $t°C$ (calculated with $t = -8\ 1 + 1\ 06h - 0\ 05h^2$)	(4) Obs. −calc
1	4 9	−4.0	−4 1	+0.1
2	6 7	−3 1	−3 2	+0 1
3	9 0	−2 5	−2 6	+0 1
4	11 3	−2 9	−2 5	−0 4
5	19 0	−5 8	−5 9	+0 1
Σ(+)				+0 4
Σ(−)				−0.4

In cases like this, it is not easy to determine the exact nature of the curved line, so the use of a quadratic equation is recommended. The form is.

$$x = A + By + Cy^2$$

where y is the independent variable.

According to Table 15, five pairs of values can be used for computing the coefficients A, B, C.

Even in this problem, the use of the method of least squares can be avoided if the observed values lie in a smoothed curve, as is seen in the actual example.   Then a good approximation can often be reached by choosing only three equations for computing the three constants.   Naturally, points on the curve which charac-

---

in verschiedenen Seehohen" (Variability and Depth of Snow on the Ground in Different Altitudes, with an English summary), *Gerlands Beitrage zur Geophysik*, vol. LI, 1937, pp 100–128. In the East Alps, the variation of the depth of snow on the ground with altitude can be well represented by hyperbolas from November through March   In April, and this is of great methodological interest, the hyperbola must be replaced by a parabola, which considers the melting of the snow cover in the lower levels in spring, on the one hand, and a further increase of the depths of snow in the high levels, on the other hand.   The hyperbola reproduces the relations between depth of snow and altitude as far as in every level an *increase* of the depths takes place

[6] Example from V Conrad, *Klimatographie von Karnten* (Wien, 1913)

terize the trend as well as possible should be selected from the graph.

In the present example, the following equations were chosen·

$$-4.0 = A + 4.9\,B + 24\,C$$
$$-2.5 = A + 9.0\,B + 81\,C$$
$$-5.8 = A + 19.0\,B + 361\,C$$

These are the pairs of values nos. 1, 3, 5. The equations are easily solved. It is necessary to change the term A by only 0.1 C° to equalize the sum of positive and negative differences "obs. — calc." It follows the equation:

$$t = -\,8.0 + 1.05\,h - 0.05\,h^2 \;(°C)$$

where t means the temperature, h the height in hectometers. The constants are:

$$A = -\,8.0; \qquad B = 1.05; \qquad C = -\,0.05$$

The differences "obs. — calc." show clearly that, in general, in a climatological problem greater accuracy is unnecessary; another choice of intervals, another choice of stations, etc., would cause slight changes of the vertical temperature distribution which are of the same order of magnitude as the differences "obs. — calc." Therefore the application of more exact methods would be a waste of time.

The advantages of representing variations of elements by analytical equations are obvious. The equations permit an objective interpolation, based on the totality of observations.

IV. 4. a. *Level of the Turning Point of the Curve*

As a supplement to the last problems, two applications of the representation of observations by means of analytical equations are given. From the observations (see Fig. 14), it is clear that the temperature increases in the lower layers with height (inversion), and above a certain level the normal decrease of temperature occurs.

Question: At what altitude is the turning point of the temperature-height curve located?

At the extremes, the differential quotient of a function becomes zero.

$$\frac{dt}{dh} = 1.05 - 2 \times 0.05\, h = 1.05 - 0.1\, h$$

$$\left(\frac{dt}{dh}\right)_{t=max} = 1.05 - 0.1\, h = 0$$

and hence $(h)_{t=max} = 10.5$ hectometers, or $= 3445$ ft.
This level is the average top-level of the inversion in winter in the mountain valley in question.

IV. 4. b. *The Standard Distribution*

As was said before, the analytical equation permits full interpolation of the element for every value of the independent variable. In the present example, the average winter temperature can be given for every desired altitude which lies between the extreme levels of the places of which temperatures are used in the computation.

For some purposes, it is very important to know the average temperatures at the consecutive levels, for instance, 10-meter or 100-meter levels, above the bottom of the valley.

A clipping from such a table calculated from the equation

$$t = -8.0 + 1.05\, h - 0.05\, h^2$$

is given in Table 16. It indicates the average temperatures every

TABLE 16   CLIPPING FROM A TABLE OF AVERAGE DISTRIBUTION OF TEMPERATURE ABOVE THE BOTTOM OF A MOUNTAIN VALLEY IN JANUARY

Altitude	00	10	20	30	40	etc
meters			Centigrade degrees			
400	−4 6	−4.5	−4 5	−4 4	−4.4	
500	−4 0	−3 9	−3 9	−3.8	−3 8	
600	−3 5	−3.5	−3.4	−3 4	−3 3	
etc						

10 m. The content of such a table represents the *standard distribution* of temperature with height, in the region in question.

This representation of the distribution of temperature with height is naturally only one example out of many. The method can and should be applied to any climatological element which is correlated with a climatic factor, or with another climatic element. (See XIV, 4 and 5.)

IV. 5. SMOOTHING OF NUMERICAL SERIES

This procedure is usual in climatological practice, if the averaged variation of an element—for instance, an average annual course—presents an irregular aspect. *One assumes that the irregularities are of incidental nature and would diminish in proportion to the length of the available period.*

Different methods are applicable. Some, of greater importance for the climatologist, are mentioned below, in each of which a series of the following ten numbers will be used for a numerical example: 12, 31, 9, 4, 11, 10, 15, 7, 11, 13.

1) *Overlapping sums:*

(a) of three consecutive elements according to the formula

$$\bar{a}_i = (a_{i-1} + a_i + a_{i+1}) : 3$$

where a_i means any element in the series, the subscript $(i-1)$ stands for the preceding, $(i+1)$ for the following, element. $\bar{a}_i$ is the smoothed value replacing the original a_i. Example.

$$a_1, \ a_2, \ a_3, \ a_4, \ a_5, \ a_6, \ a_7, \ a_8, \ a_9, \ a_{10}$$
$$12, \ 31, \ 9, \ 4, \ 11, \ 10, \ 15, \ 7, \ 11, \ 13$$

The series may be considered as cyclic (for instance, an annual course). Then, the smoothing procedure begins with the last item and ends with the first, and the smoothed series reads:

$$\bar{a}_1 = (a_{10} + a_1 + a_2) : 3 = (13 + 12 + 31) \cdot 3 = 19$$
$$\bar{a}_2 = (a_1 + a_2 + a_3) : 3 = (12 + 31 + 9) : 3 = 17$$
$$\cdot \qquad \cdot \qquad \cdot \qquad \cdot \qquad \cdot \qquad \cdot \qquad \cdot$$
$$\bar{a}_{10} = (a_9 + a_{10} + a_1) : 3 = (11 + 13 + 12) : 3 = 12$$

so, finally, we have the sequence of smoothed values:

$$19, \ 17, \ 15, \ 8, \ 8, \ 12, \ 11, \ 11, \ 10, \ 12$$

(b) The number of overlapping consecutive items can be increased arbitrarily, considering the total number of items of the series. For simple reasons, odd numbers should be preferred. As a second example, the formula for 5 consecutive items is given.

$$\bar{a}_i = (a_{i-2} + a_{i-1} + a_i + a_{i+1} + a_{i+2})$$

The item $\bar{a}_1$ of the series (1a) is therefore:

$$\bar{a}_1 = (a_9 + a_{10} + a_1 + a_2 + a_3) : 5$$
$$= (11 + 13 + 12 + 31 + 9) : 5 = 15$$

$$\bar{a}_2 = (a_{10} + a_1 + a_2 + a_3 + a_4) : 5$$
$$= (13 + 12 + 31 + 9 + 4) : 5 = 14 \qquad \text{etc.}$$

2) *Weighted overlapping sums:*

(a) $$\bar{a}_\imath = (a_{\imath-1} + 2a_\imath + a_{\imath+1}) : 4$$

Numerically.

$$\bar{a}_1 = (a_{10} + 2a_1 + a_2) \ 4$$
$$= (13 + 2 \times 12 + 31) : 4 = 17 \qquad \text{etc.}$$

(b) $$\bar{a}_\imath = \left. \begin{array}{l} (a_{\imath-4} + 2a_{\imath-3} + 3a_{\imath-2} + 4a_{\imath-1} + 5a_\imath + \\ a_{\imath+4} + 2a_{\imath+3} + 3a_{\imath+2} + 4a_{\imath+1}) \end{array} \right\} : 25$$

Numerically:

$$\bar{a}_1 = \left. \begin{array}{l} (a_7 + 2a_8 + 3a_9 + 4a_{10} + 5a_1 + \\ a_5 + 2a_4 + 3a_3 + 4a_2) \end{array} \right\} : 25$$

$$\bar{a}_1 = (15 + 14 + 33 + 52 + 60 + $$
$$11 + 8 + 27 + 124) = 14 \quad \text{etc.}$$

(c) A more individual weighting is obtained by using the binomial coefficients as weights for the consecutive items. Formula 2a represents the simplest case, the binomial coefficients of a squared binom:

i.e., 1, 2, 1

The binomial coefficients for the 4th power are: 1, 4, 6, 4, 1; and those for the 6th power: 1, 6, 15, 20, 15, 6, 1.

Therefore, the smoothing formula, corresponding to the 4th power, is:

$$\bar{a}_\imath = (a_{\imath-2} + 4a_{\imath-1} + 6a_\imath + 4a_{\imath+1} + a_{\imath+2}) . 16$$

The first term of our numerical series when smoothed is:

$$\bar{a}_1 = (a_9 + 4a_{10} + 6a_1 + 4a_2 + a_3) . 16$$
$$= (11 + 52 + 72 + 124 + 9) : 16 = 17$$

Numerical series should be smoothed only if absolutely necessary. The smoothing procedure can efface essential characteristics of the series and lead to false conclusions.

CHAPTER V

HARMONIC ANALYSIS

V. 1. THE ANALYSIS

THE HARMONIC ANALYSIS is one of the methods for describing periodic phenomena. The latter are indicated by a variation in which the dependent variable shows a repetition at equal intervals of the other variant. A great number of climatological elements reach one extreme value at about low sun and the opposite extreme at about high sun. The periodical variation of the altitude of the sun is the physical cause of the periodical annual variation of these elements. Their daily course frequently offers a complex problem. The physical conditions during the day are different from those during the night. A sound physical standpoint should never be neglected in investigations of periodical phenomena

The harmonic analysis makes use of Fourier's series. We start from the following formula, which is rather common in meteorological and climatological investigations:

$$y = a_o + a_1 \sin (x + A_1) + a_2 \sin (2x + A_2) + \cdots$$
$$+ a_i \sin (ix + A_i) + \cdots$$

where a_o is the arithmetic mean of the observations and $a_1, a_2, \cdots a_i$ are amplitudes (half ranges) of the superimposed waves. A_1, $A_2, \cdots A_i$ are called *phase angles*, determining the times at which the extremes occur. The meaning of the equation mentioned is easy to understand by means of the graph of Figure 15.

The symbol x represents the time angle. It is computed from

$$x = \frac{360°}{P}$$

if P denotes the length of the period investigated. For example, if $P = 12$ months, then

$$x = \frac{360°}{12} = 30°, \qquad 2x = 60°, \qquad \text{etc.}$$

70

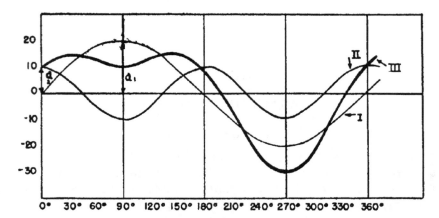

FIG. 15   Example of superposition of two waves

$$I = a_1 \sin \ (x + A_1), \qquad A_1 = \ 0°$$
$$II = a_2 \sin (2x + A_2), \qquad A_2 = 90°$$
$$III = \text{sum of the waves I and II}$$

If $P = 24$ hours

$$x = \frac{360°}{24} = 15° \qquad \text{etc.}$$

Further, it is easy to see that

$$y = a_o + a_1 \sin x \cos A_1 + a_2 \sin 2x \cos A_2 + \cdots$$
$$+ \ a_1 \cos x \sin A_1 + a_2 \cos 2x \sin A_2 + \cdots$$

We substitute.

$$p_1 = a_1 \sin A_1; \qquad p_2 = a_2 \sin A_2 \cdot$$
$$q_1 = a_1 \cos A_1; \qquad q_2 = a_2 \cos A_2 \cdots$$

Then,

$$y = a_o + p_1 \cos x + p_2 \cos 2x + \cdots + q_1 \sin x + q_2 \sin 2x + \cdots$$

$u_o, u_1, \cdots u, \cdots u_{n-1}$ may be the deviations from the arithmetical mean and are inserted into the equation in place of y.

Then we get n equations from which the constants p_1, q_1, p_2, $q_2 \cdots$ can be calculated.

If the number of observations is greater than that of the constants (the common case), the method of least squares is to be applied. This method yields the following equations for the constants:

$$p_1 = 2/n[u_o \cos 0° + u_1 \cos x + u_2 \cos 2x + \cdots u_{n-1} \cos (n-1)x]$$

$$q_1 = 2/n[u_o \sin 0° + u_1 \sin x + u_2 \sin 2x + \cdots u_{n-1} \sin (n-1)x]$$

$$p_2 = 2/n[u_o \cos 0° + u_1 \cos 2x + u_2 \cos 4x + \cdots u_{n-1} \cos 2(n-1)x]$$

$$q_2 = 2/n[u_o \sin 0° + u_1 \sin 2x + u_2 \sin 4x + \cdots u_{n-1} \sin 2(n-1)x]$$

$$\cdot \quad \cdot \quad \cdot \quad \cdot \quad \cdot \quad \cdot \quad \cdot \quad \cdot \quad \cdot \quad \cdot \quad \cdot \quad \cdot$$

$$p_k = 2/n[u_o \cos 0° + u_1 \cos kx + u_2 \cos 2kx + \cdots u_{n-1} \cos (n-1)kx]$$

$$q_k = 2/n[u_o \sin 0° + u_1 \sin kx + u_2 \sin 2kx + \cdots u_{n-1} \sin (n-1)kx]$$

According to the definitions of p_k and q_k,

$$\frac{p_k}{q_k} = \tan A_k \quad \text{and} \quad \frac{p_k}{\sin A_k} = a_k$$

Obviously, also, the following solutions are valid:

$$p_k{}^2 + q_k{}^2 = a_k{}^2, \quad \text{or} \quad a_k = \sqrt{p_k{}^2 + q_k{}^2}$$

L. W. Pollak [1] introduced an excellent procedure for evaluating the equations mentioned, which is so clear and simple that even those without special mathematical training are able to make the necessary calculations.

In the following, therefore, Pollak's system is illustrated. For lack of space, only two special problems, frequently used in climatology, can be discussed:

    1) 12 equidistant observations (average values)
    2) 24 equidistant observations

Case 1 usually represents the annual course of an element; case 2, the diurnal course. There is, however, no difficulty in calculating the constants of a diurnal course with 12 two-hourly

[1] L W Pollak, "Handweiser zur harmonischen Analyse," *Prager Geophysikalische Studien*, vol II (Prague, 1928); L. W Pollak, *Rechentafeln zur harmonischen Analyse*, (Leipzig, 1926). I learn from a letter that Dr. Pollak is extending the *Rechentafeln* up to 100 ordinates The first edition reached 40 ordinates.

values and in evaluating an annual variation by means of 24 half-monthly means, if greater accuracy is required.

Generally, the monthly means are related—as was said above—to the middle day of the month: e.g. the 14th, 15th, or 16th day of

TABLE 17   INSTRUCTIONS FOR CALCULATING THE CONSTANTS FOR 12 EQUIDISTANT VALUES YIELDING THE FIRST, SECOND, AND THIRD TERM OF FOURIER'S SERIES
(After L W Pollak)

s	p_1	q_1	p_2	q_2	p_3	q_3
0	1	0	1	0	1	0
1	0 866	0 5	0 5	0 866	0	1
2	0 5	0 866	−0 5	0 866	−1	0
3	0	1	−1	0	0	−1
4	−0 5	0 866	−0 5	−0.866	1	0
5	−0 866	0 5	0 5	−0 866	0	1
6	−1	0	1	0	−1	0
7	−0 866	−0 5	0 5	0 866	0	−1
8	−0 5	−0 866	−0 5	0 866	1	0
9	0	−1	−1	0	0	1
10	0 5	−0 866	−0 5	−0 866	−1	0
11	0 866	−0 5	0 5	−0 866	0	−1

TABLE 18   THE SAME AS IN TABLE 17, BUT FOR 24 EQUIDISTANT VALUES
(After L W Pollak)

s	p_1	q_1	p_2	q_2	p_3	q_3
0	1	0	1	0	1	0
1	0 966	0.259	0 866	0 5	0 707	0 707
2	0 866	0 5	0 5	0 866	0	1
3	0 707	0 707	0	1	−0 707	0 707
4	0.5	0.866	−0 5	0 866	−1	0
5	0 259	0 966	−0 866	0 5	−0 707	−0 707
6	0	1	−1	0	0	−1
7	−0 259	0 966	−0 866	−0.5	0 707	−0 707
8	−0 5	0 866	−0 5	−0 866	1	0
9	−0 707	0 707	0	−1	0 707	0 707
10	−0 866	0 5	0 5	−0 866	0	1
11	−0 966	0 259	0 866	−0.5	−0 707	0 707
12	−1	0	1	0	−1	0
13	−0 966	−0 259	0.866	0 5	−0.707	−0 707
14	−0 866	−0 5	0 5	0 866	0	−1
15	−0 707	−0 707	0	1	0 707	−0 707
16	−0 5	−0 866	−0.5	0 866	1	0
17	−0 259	−0 966	−0 866	0 5	0 707	0 707
18	0	−1	−1	0	0	1
19	0 259	−0 966	−0 866	−0 5	−0 707	0.707
20	0 5	−0 866	−0.5	−0 866	−1	0
21	0.707	−0 707	0	−1	−0 707	−0 707
22	0 866	−0 5	0 5	−0 866	0	−1
23	0.966	−0.259	0.866	−0 5	0 707	−0 707

the month, depending on the length of the respective months. Hourly means are related to the 30th minute of the hour.

Tables 17 and 18 contain instructions for calculating the constants

$$p_1, q_1, p_2, q_2, p_3, q_3,$$

and from these,

$$a_1, a_2, a_3 \quad \text{and} \quad A_1, A_2, A_3$$

The first column of each table contains the number of the item. For example: the mean of May has the number $i = 4$, because January has the number zero. The value of 5 A.M. has number $i = 5$; if the series begins with midnight $= 0$ hour.

In columns $p_1, q_1. p_2, q_2 \cdots$ the factors by which the respective items (see Column i) have to be multiplied, are indicated.

The columns p_1, q_1, etc., are totaled (naturally, with regard to sign).[2] The sums multiplied by $2/n$ (that is, $1/6$ for 12 values, and $1/12$ for 24 values) yield the constants p_1, q_1, etc. The multiplications are done with the aid of a slide rule, a multiplication table, or a calculating machine. The best expedient is Pollak's *Rechentafeln*, which are indispensable if such calculations are made on greater scale, or if other lengths of period are considered. In these calculations, however, one should shun a purely arithmetical accuracy not guaranteed by the exactness of the observations. Generally, results to three places of decimals are ample.

If the constants p and q are calculated, the amplitude a and the phase angle A, are easily computed with the formulae given above.

Although the problem is exceedingly simple, blunders sometimes happen in determining the true phase angle. The diagram (Fig. 16), helps to avoid such mistakes. Possible combinations of the signs of p and q for which the phase angle is to be determined, are:

a) $\quad + p + q;$
b) $\quad + p - q;$
c) $\quad - p - q;$
d) $\quad - p + q$

---

[2] The deviations $u_0, u_1, \cdot, u_{n-1}$ are multiplied with the factors of each of the columns $p_1, q_1; p_2, q_2, \cdots$ These series of products are entered into a new blank, with the same headings $p_1, q_1; p_2, q_2; \cdots$. The column p_1 of the new blank contains, thus, the following products (in the case of Table 17).

$$1 \quad \times u_0$$
$$0.866 \times u_1$$
$$\cdots \cdots \cdots \cdots$$
$$0.866 \times u_{n-1} \quad \text{etc.}$$

These columns $p_1, q_1; p_2, q_2; \cdots$ are then totaled

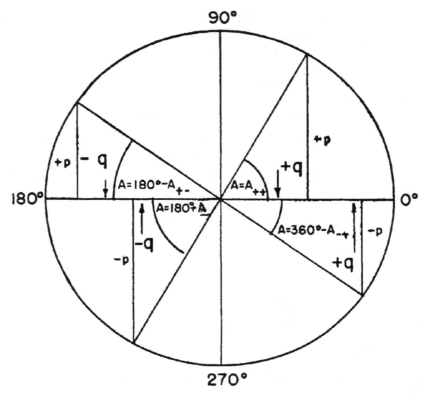

FIG. 16. Diagram for determining the true phase angle

From these combinations, we calculate directly the following angles:

a) A_{++};   b) A_{+-};   c) A_{--};   d) A_{-+}

Then the true phase angle A is:

a) $A = A_{++}$ A in the first quadrant
b) $A = 180 - A_{+-}$ A in the second quadrant
c) $A = 180 + A_{--}$ A in the third quadrant
d) $A = 360 - A_{-+}$ A in the fourth quadrant

The following example,[3] Table 19 shows each step of the calculation, so that a few explanatory remarks are sufficient.

[3] Taken from V. Conrad, "Anomalien und Isanomalen der Sonnenscheindauer in den oesterreichischen Alpen," *Beihefte Jahrbuch d Zentralanstalt f. Meteorologie* (Wien, 1938); V. Conrad, "Die Komponenten der Jahresschwankung der Sonnenscheindauer," *Helvetia Physica Acta*, vol XII (1939), p. 38.

TABLE 19. HARMONIC ANALYSIS OF THE ANNUAL COURSE OF THE RELATIVE SUNSHINE DURATION (S in %) AT AN AVERAGE LEVEL OF 1000 FEET IN A MOUNTAINOUS COUNTRY

(d_i = deviations)

(1) Month	(2) S(%)	(3) i	(4) d_i	(5) p_1	(6) q_1	(7) p_2	(8) q_2
I	25	0	−19	− 19	0	−19	0
II	39	1	− 5	− 4.330	− 2 5	− 2.5	− 4 330
III	47	2	+ 3	+ 1.5	+ 2.598	− 1 5	+ 2 598
IV	45	3	+ 1	0	+ 1	− 1	0
V	53	4	+ 9	− 4 5	+ 7 794	− 4.5	− 7 794
VI	58	5	+14	− 12 124	+ 7	+ 7	−12 124
VII	62	6	+18	− 18	0	+18	0
VIII	58	7	+14	− 12.124	− 7	+ 7	+12.124
IX	54	8	+10	− 5	− 8 660	− 5	+ 8 660
X	41	9	− 3	0	+ 3	+ 3	0
XI	29	10	−15	− 7.5	+12 990	+ 7.5	+12 990
XII	19	11	−25	− 21 650	+12.5	−12 5	+21 650
mean (a_0)	44			− 102 728	+28 722	− 3 5	+33.774

$p_1 = − 17.121$ lg $p_1 = 1 2335$ $p_2 = − 0 583$ lg $p_2 = 9 7657$

$q_1 = + 4.787$ lg $q_1 = 0 6801$ $q_2 = + 5 629$ lg $q_2 = 0 7504$

$(A_1) = 74°22'$ lg tan $A_1 = 0 5534$ $(A_2) = 5°55'$ lg tan $A_2 = 9 0153$

$a_1 = 17 8\%$ lg $p_1 = 1.2335$ $a_2 = 5 7\%$ lg $p_2 = 9.7657$

$A_1 = 285°38'$ lg sin $A_1 = 9 9836$ $A_2 = 354°5'$ lg sin $A_2 = 9.0132$

lg $a_1 = 1.2499$ lg $a_2 = 0.7525$

$$S = 44 + 17\,8 \sin (x + 286°) + 5.7 \sin (2x + 354°)$$

Column 2 contains the original numbers, the periodical trend of which is to be analyzed. Columns 5 to 8 contain the products of the consecutive deviations (see column 4), and of the values of the trigonometrical functions of the respective angles as indicated in Table 17.

V. 2. EVALUATION OF THE EQUATION

Table 20 gives a pattern of the evaluation of Fourier's series:

$$S = 44 + 17.8 \sin (x + 286°) + 5.7 \sin (2x + 354°)$$

This evaluation is to be recommended, since it is the only check on the correctness of the analysis. The graph, Figure 17, furnishes a good illustration of the agreement of observation and calculation. The example shows, too, that the combination of only two waves is sufficient to give a good approximation with the rather complex

TABLE 20. EVALUATION OF EQUATION S IN TABLE 19

(1) Month	(2) $x = °$	(3) $(x + A_1)°$	(4) $2x + A_2$	(5) 18 sin $(x + A_1)$	(6) 6 sin $(2x + A_2)$	(7) C.	(8) O	(9) O − C
Jan.	0	−74°22′	− 5°55′	−17 1	−0.6	−18	−19	−1
Feb.	30	−44°22′	+54°05′	−12.4	+4 6	− 8	− 5	+3
March	60	−14°22′	+65°55′	− 4.4	+5 2	+ 1	+ 3	+2
April	90	+15°38′	+ 5°55′	+ 4 8	+0 6	+ 5	+ 1	−4
May	120	+45°38′	−54°05′	+12.7	−4.6	+ 8	+ 9	+1
June	150	+75°38′	−65°55′	+17 2	−5.2	+12	+14	+2
July	180	+74°22′	− 5°55′	+17 1	−0 6	+17	+18	+1
Aug.	210	+44°22′	+54°05′	+12 4	+4 6	+17	+14	−3
Sept.	240	+14°22′	+65°55′	+ 4 4	+5.2	+10	+10	0
Oct.	270	−15°38′	+ 5°55′	− 4 8	+0 6	− 4	− 3	+1
Nov.	300	−45°38′	−54°05′	−12 7	−4 6	−17	−15	+2
Dec.	330	−75°38′	−65°55′	−17 2	−5 2	−22	−25	−3

$$\Sigma(+) = 12$$
$$\Sigma(-) = 11$$

C. = Calculated deviations
O. = Observed deviations

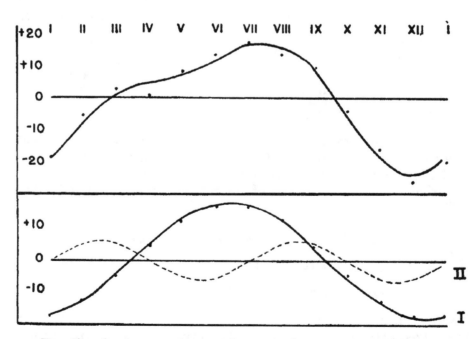

FIG 17. At the top the annual course of the relative duration of sun-shine in a given average level of a mountainous region.   Full line: calculated. Dots· observed.

At the bottom: the two constituents of the calculated curve.

observed curve.   As far as the approximation is concerned, attention should be directed to the convergence of the resulting series.

The ratio $\dfrac{a_2}{a_1} = 1/3$, and $\dfrac{a_3}{a_1} = 1/9$ (a_3 not included in Table 19), so that here a good convergence exists.   Thus, the series can be cut off at the second term, in the example of Table 19.[4]

V.   3. Relative Amplitudes

The amplitudes $a_1, a_2, a_3, \cdots$ are given in the units of the deviations.   The meaning of the amplitudes is dependent upon the particular arithmetical mean.   For purposes of comparison, the amplitudes of the different terms have to be freed from the influence of the unit and that of the arithmetical mean: to do this, the amplitudes are divided by the arithmetical mean.

$$a_{R,1} = \frac{a_1}{a_o} \qquad a_{R,2} = \frac{a_2}{a_o}$$

In the example (Table 19):

$$a_{R,1} = 0.40 \qquad a_{R,2} = 0.13$$

Often it is advantageous to give the relative amplitudes in per cent of the arithmetic mean.   Then

$$a_{R,1} = 40\%, \qquad a_{R,2} = 13\%$$

V.   5. Times of the Extremes

The times of the extremes of the single constituent waves are easily calculated from the equation:

$$kx + A_k = 90° \qquad \text{or} \qquad 450° \text{ respectively,}$$

[4] D Brunt (*Meteorological Magazine*, 1937, p 268), gives a criterion as to whether or not it pays to calculate further terms of the Fourier's series.   Brunt's formula reads:

$$\sigma'^2 = \sigma^2 - \tfrac{1}{2}(a_1^2 + a_2^2),$$

where σ^2 means the *variance* of the value (d_t, column 4 in Table 19) subject to the analysis In the example

$$\sigma^2 = \frac{\Sigma d_t^2}{n} = 179$$

$$\tfrac{1}{2}(a_1^2 + a_2^2) = 175$$

Therefore

$$\sigma'^2 = 4$$

This means that σ'^2 is only 2% of σ^2.   Hence the criterion indicates that further terms of the series can be neglected

where k equals the number of the term. Example from Table 19:

$$\left.\begin{array}{l} A_1 = 286° \\ k = 1 \end{array}\right\} \text{Time of maximum} = x + A_1 = x + 286° = 450° \quad \text{or}$$

$$x_{max} = 164°$$
$$x_{min} = 164° + 180° = 344°$$

Since 360 degrees correspond to an average year of 365.25 days, the angle-degrees have to be multiplied by 1.0145, if the conversion into days is required. The date is obtained by adding 16 to the product. This is clear, because the mean values of the elements are related to the midpoint of the month (generally January).

Example: If $x = 164°$, $1.0145 \times 164 = 166$ days have been passed between the middle of January and the time of the maximum.[5] Therefore the maximum of the whole-year-wave occurs on the $166 + 16 = 182$nd day of the year: that is, July 1.[6]

It goes without saying that the starting date can be chosen arbitrarily. The hydrologist, for instance, begins his year on July first, October first, November first, or any other date representing the beginning of the chief rainy season.

The problem of calculating the time of the extremes directly from Fourier's series with two terms was simply solved nearly a century ago. The solution gives a good approximation.[7] Generally, it is sufficient to take the times of the extremes from a good graph. In our example, the curve representing the annual course of the relative duration of sunshine (Fig. 17) indicates, approximately, August 1 for the maximum and December 16 for the minimum.

Harmonic analysis permits the interpolation of missing or dubious values of a periodical series and is an excellent means of investigation in many problems of periodicity. The example of Table 19 is taken, for instance, from an investigation in the variation of the annual course of the duration of sunshine with altitude. By means of the harmonic analysis, it was possible to puzzle out the complex problem and to show—among other results—that the amplitude of the second harmonic term is invariant with height.[8] In order to illustrate the high research value of the harmonic an-

---

[5] It is sufficient to increase the number of degrees by $1\frac{1}{2}\%$ [6] See Appendix IV.

[7] Hugo Meyer, *Anleitung zur Bearbeitung meteorologischer Beobachtungen* (Berlin, 1891), p. 38 ff.

[8] This result is important in so far as it indicates that, against every expectation, the half-year wave is the main wave upon which the whole-year wave of very variable amplitude and phase angle is superposed.

alysis, attention is drawn to Hann's excellent studies of the daily course of air pressure, as also to the numerous investigations which deal with the idea of Arthur Schuster's "expectancy," and to the methods of L. W. Pollak and Sir Gilbert Walker.

Even if the climatologist is not compelled *de facto* to use the harmonic analysis for his own work and investigations, he must understand this method, so often applied in scientific literature.

PART II

REPRESENTATION OF CHARACTERISTIC FEATURES OF DIFFERENT ELEMENTS

I T IS NEITHER desirable nor possible to give detailed instructions as to how to handle each element and its characteristics. The first part of this book has offered the mathematical means of descriptive statistical characteristics for every series of numbers, independent of the element. The only restriction—already mentioned—was that here scalar quantities, not vectorial, were considered.

Many of the common characteristics have been discussed previously. A partial summary of the most usual characteristics is given in the forms of climatic tables in the Appendix. The following sections deal with a selection only of important and less well-known characteristics.

CHAPTER VI

TEMPERATURE

VI. 1. Hours of Observation. Reduction to the "True Mean"

T HE INTERNATIONAL REGISTER assumes that the elements, with the exception of precipitation, are observed three times a day. This system is used in most of the European countries, in Russia, from the Baltic Sea to the east Asiatic coast, and in some other networks of climatological stations. The method of three daily observations is therefore important for the climatographies of at least two continents. The observations are so distributed over the day that one is made in the morning, one in the afternoon, and the third in the evening.

For an easier understanding, Figure 18 represents a daily course of temperature derived from 50 years' thermograph records and related to the middle day of the month. The average temperature is characterized by the ordinate of that rectangle which is equal to the area enclosed by the curve (top), the abscissa (bottom), in Figure 18 located at 14°C, and the limiting ordinates (sides). Generally this area is measured by a planimeter. In the case of a smooth curve, such as that in Figure 18, one can estimate the

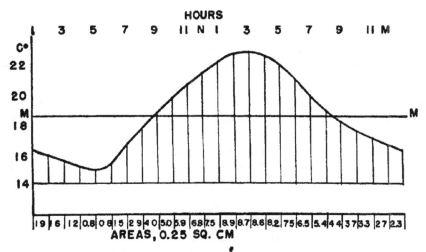

FIG. 18.   The daily course of temperature at a place in
Central Europe, in August (C°)

areas between the ordinates of two consecutive hours; the sizes of these areas can be read from the numbers on the X-axis. The unit is 0.25 sq cm. The sum of these areas divided by the number of intervals (i.e., 24) gives the height of the average ordinate. In the present case, 4.57°C results.

Therefore the average temperature required is 4.57 + 14.00 = 18.57°C. The method of numerical integration is correct but complicated.

TABLE 21   AVERAGE DAILY COURSE OF TEMPERATURE AT
A PLACE IN CENTRAL EUROPE (AUGUST)

	1	2	3	4	5	°C 6	7	8	9	10	11	12
A.M.	16 1	15.8	15 4	15.1	14 8	15 0	16 1	17 4	18.5	19.4	20 3	21.1
P.M.	21.9	22 6	22 7	22.5	21.9	21.0	19 8	18.8	18.1	17.5	17 0	16.6

arithmetic mean = 18 56 °C

We can arrive also at a practically identical result by a very simple computation. The arithmetic mean of the 24-hourly temperatures (see Table 21) yields 18.56°C.

Thus the arithmetic mean of the 24-hourly temperatures is defined as the *true mean* temperature.[1] This definition is valid for all elements which are continuously recorded or observed at each hour, except vectorial values.

We return to the international form of the climatological register and observations at certain fixed hours of the day. The aim of these repeated observations during the day is to learn something about the daily variation of temperature (or another element), on the one hand; and, on the other hand, to figure out the average temperature of an individual day, and the true means for the months and for the year.

Two statistical results are immediately derived from the three (two in some networks) daily observations: (1) The average of the daily observations is defined as the daily mean (valid for all elements except wind direction). (2) The temperatures of each observation hour are averaged over the month. Monthly average temperatures for the fixed hours result.

A simple combination of these average observational hours should yield the true monthly mean. A difficulty exists in so far as the form of the daily course is variable from region to region, and in the course of the year.

Before we continue this discussion, it is necessary to give some numerical examples of observational hours in the United States and elsewhere.

The hours which are highly recommended by the International Meteorological Committee [2] and used in many European meteorological networks are:

7 A.M., 2 P.M., 9 P.M.

[1] The usual nomenclature is *true daily mean*. This denomination is appropriate as long as 24-hourly temperatures are concerned. Then the expression is valid for any single day as well as for the average of the month. It is another matter with observations at fixed hours. We then have to discriminate between: (a) the daily mean and (b) the mean of the monthly averages of the temperatures at fixed hours. In case (a), we speak only of a *daily mean*, which will be discussed later on. Only from the monthly average temperatures at the fixed hours can a combined and corrected value be derived which shows an optimum of agreement with the 24-hour mean. This approximation is called *true mean* in the following discussion.

[2] *International Meteorological Codex*, p. 24.

Other recommended hours are:

> 6 A.M., 2 P.M., 10 P.M.
> 7 A.M., 2 P.M., 10 P.M.
> 7 A.M., 1 P.M., 9 P.M.
> 8 A.M., 2 P.M., 8 P.M.[3]

In the United States the hours of observations shown in Table 22 were used.[4]

TABLE 22   OBSERVATION-HOURS IN THE UNITED STATES (AFTER FRANK BIGELOW) [a]

1870, Nov to 1872, Aug 24,	7 35 A.M , 4 35 P.M , 11.35 P.M. WMT[b]
1872, Aug 24, to 1879, Oct 31,	7.35 A M , 4.35 P M , 11 00 P M WMT
1879, Nov 1, to 1886, Dec 31,	7.00 A M , 3.00 P M., 11 00 P M WMT
1887, Jan 1, to 1888, June 30,	7 00 A M , 3 00 P M , 10 00 P M 75th MT
1888, July 1, till about 1930	8 00 A M , 8 00 P M 75th MT
1930 to 1936,	7 00 A M., 7 00 P M.
From 1936,	7 00 A M , 1.00 P M., 7 00 P M., 1 00 A.M.

[a] Dr C F Brooks has been so kind as to bring down to date Bigelow's data
[b] WMT = Mean time of the meridian of the observatory at Washington, D C.
75th MT = Mean time of the meridian 75°W = Eastern Standard Time
WMT — 75th MT = 8 3 minutes

Table 23 contains examples of the annual course of the *difference* between the temperature obtained from the combination of the observational hours (t_o) and the true mean (derived from 24-hourly observations, (t_t)). The difference

$$c = t_t - t_o$$

is called *correction to the "true mean.'*   It is

$$t_t = t_o + c$$

This means that the correction c is to be added to t_o without changing its sign.

[3] There is also a series of only two observational hours, recommended by the IMC:

> 8 A.M., 8 P M.
> 9 A.M , 9 P M
> 10 A M., 10 P M.

The averages of the temperatures at these hours do not show great differences from the true mean but do not indicate anything about the daily variation of temperature.

[4] According to the investigations in this problem by Frank H. Bigelow, "Report on the Temperatures and Vapor Tensions of the United States," U S Dept of Agriculture, Weather Bureau, *Bulletin S* (Washington, 1909) See also Alexander McAdie, *Mean Temperatures and Their Corrections in the United States* (U. S War Dept., Washington, 1891).

TABLE 23   THE ANNUAL COURSE OF "c" (F°) FOR
DIFFERENT HOURS OF OBSERVATION

Hours	Buffalo, N. Y (42.9°N, 78 9°W, 770 ft)			Central Europe	
	7 A M 3 P M 11 P M	8 A M 8 P M	max min	7 A M. 2 P M 9 P M	7 A M 2 P M 2 × 9 P M
Jan.	0 0	+0 4	+0 4	−0 9	+0 4
Feb.	0 0	+0 6	+0 4	−0 5	−0 2
Mar.	+0.1	+0 8	+0 2	−0.3	+0.1
Apr.	+0 2	+0 5	−0 2	−0 4	−0 1
May	+0.3	+0.2	+0.1	−0.7	−0 3
June	+0 3	0.0	+0.3	−0 9	−0 5
July	+0.3	0 0	+0 3	−0.9	−0 5
Aug	+0 3	+0 4	+0 3	−0 1	+0.3
Sept	+0 4	+0 8	+0 4	−0.1	+0.3
Oct	+0 3	+1.2	+0 5	−0 4	0 0
Nov	+0 1	+0 6	+0 3	−0 8	−0.3
Dec.	0.0	+0 4	+0 5	−1.0	−0 7
Year	+0.2	+0 5	+0 3	−0 6	−0 2
$\Sigma|c|$	2 3	5.9	3 9	7 0	3 7
A R	0.4	1.2	0.7	0 9	1 1
A Rx$\Sigma|c|$	0 9	7.1	2.7	6 3	4 1

The first column shows c for three daily observations, the second for two, the third for the extremes, the fourth for three other hours, and the fifth offers a new feature. Here the observations at the three different hours are not combined in a common arithmetic mean, but in a *weighted* mean. Double weight is given to the average temperature at 9 P.M. Later on, we shall discuss this method. The question now to be answered is, which hours and which combinations of observations are most advantageous.

There are two requirements:

1) The sum of the absolute values of the monthly corrections should be as small as possible.
2) The annual variation of the corrections should be as small as possible.

The two requirements can be examined by simple criteria as follows: the third line from the bottom in Table 23 shows the sums of the monthly corrections, disregarding the signs ($\Sigma|c|$). It is clear that the first requirement is the better fulfilled the smaller the term $\Sigma|c|$ is.

The second line from the bottom (Table 23) contains the annual range (AR) of the corrections, i.e., the difference between the largest and the smallest correction in the course of the year. It is obvious that the product of the two quantities should be as small as possible:

$$AR \times \Sigma|c| = \text{minimum}$$

The hours 7 A.M., 3 P.M., 11 P.M., are indeed by far the best choice, and the hours 8 A.M., 8 P.M., the worst of the five. We see, too, that the result from identical hours of observation can be improved by weighted averages. The third combination (7 A.M., 2 P.M., 9 P.M.) is only slightly better than the second) (8 A.M. + 8 P.M.). The weighted mean is much better than either. (See last line of Table 23.) Nevertheless, three observations a day are much more advantageous, since they give rather good information about the daily variation of temperature. That is not the case at 8 A.M. and 8 P.M., because these times are close to the hours of the daily mean.

The difference between the average of the hours of observations and the true mean can be decreased by giving different weight to the observations—as has been shown in one special example. If we assume three observations a day, then generally, the following equation holds:

$$i = pt_a + qt_b + rt_c$$

if i stands for "true mean," the subscripts a, b, c denote the different hours of observations; the letter t, the average monthly temperature at the respective hours. The letters p, q, r are coefficients which have to be computed from the observations. Obviously the condition is valid:

$$p + q + r = 1$$

If there are complete observations at the fixed hours and synchronous, continuous records of temperature, say for two Januaries, the problem is very simple.[5]

There are i_1 and i_2 from the continuous records and $t_{a,1}$, $t_{a,2}$, $t_{b,1}$, $t_{b,2}$, $t_{c,1}$, $t_{c,2}$ for the average temperatures at the observational hours in the two January months, indicated by the subscripts 1, 2.

[5] The once chosen hours and the resulting combination of the monthly mean yields variable differences from the "true mean" in the course of the year.

Then the following equations are valid:

$$pt_{a,1} + qt_{b,1} + rt_{c,1} = l_1$$
$$pt_{a,2} + qt_{b,2} + rt_{c,2} = l_2$$
$$p + q + r = 1$$

From these three equations, the coefficients p, q, r are computed; in reality more than 2 years' observations have to be available, in which case the three coefficients can be calculated with the method of least squares.[6]

Examples of weighted arithmetic means of the temperatures at fixed hours of observation follow:

a) For the greater portion of Europe the observational hours 7 A.M., 2 P.M., 9 P.M., with double weight given the 9 P.M. temperature, are very advantageous:

$$l = (t_7 + t_2 + 2t_9) : 4 = 0.25\, t_7 + 0.25\, t_2 + 0.50\, t_9, \quad \text{or}$$
$$p = 0.25, \quad q = 0.25, \quad r = 0.50$$

b) Subtropics:

$$l = [t_7 + t_2 + t_9 - 1/10\,(t_2 - t_9)] : 3, \quad \text{or}$$
$$p = 0.33, \quad q = 0.30, \quad r = 0.37$$

c) Tropics:

$$l = [2(t_7 + t_2) + 3t_9] : 7, \quad \text{or}$$
$$p = 0.286, \quad q = 0.286, \quad r = 0.428$$

The same method is applicable if only the *extremes* of temperature are observed once a day.

This problem is of special importance. About 5500 coöperative observers in the United States make but one observation within the 24 hours of a day, and the same method is used in other countries.

For reducing the extreme temperatures to the true mean the formula used is:

$$l_t = m + k(M - m)$$

---

[6] For further information see Nils Ekholm, "Reduction of Air Temperatures at Swedish Stations to a True Mean," MWR, vol. 45 (1917), p. 58; Nils Ekholm, "Calcul de la température moyenne de l'air aux stations météorologiques Suédoises," *Appendice aux observations météorologiques Suédoises*, vol. 56, 1914 (Stockholm 1916). C. E. P. Brooks, "True Mean Temperature," MWR, vol. 49 (1921), p. 226.

For instruction for the application of the method of "least squares," see, for example, H. Arkin and R. R. Colton, *An Outline of Statistical Methods* (4th ed., New York, 1939).

where M and m stand for maximum and minimum, and k is a factor which is more or less constant for a rather large region and is valid for a certain month; k varies in the course of the year.

C. E. P. Brooks [7] studied this problem in a general manner by calculating a correction term which reduces the arithmetic mean of M and m to the true mean. His formula reads:

$$i = \frac{M + m}{2} + (a + bR + cR^2)$$

where a, b, c, are coefficients and $R = M - m$.

The correction term depends principally upon the average aperiodical range of temperature (R). It is variable in the course of the year because R has an accentuated annual course at least within the continents.

The problem is complicated, furthermore, because the coefficients a, b, c, are functions of the altitude (h) of the place above sea level. It is

$$a = -\,0.30 + 0.14\,h$$
$$b = \quad\,\, 0.00 - 0.07\,h$$
$$c = -\,0.0034 + 0.006\,h$$

The elevation h is expressed in kilometers, and the temperatures are absolute temperatures, i e., $T = t°C + 273°C$.

Example of a computation of the correction term:

Blue Hill (42.2°N. 71.1°W, 640 ft.)
July: $M = 79°F$, $m = 61°F$, $h = 640$ ft.

converted into absolute temperature and kilometer respectively:

$M = 299°$ abs., $m = 289°$ abs., $h = 0.2$ km

Therefore: $a = -\,0.27$, $b = -\,0.01$, $c = -\,0.0022$, $R = 10°C$

Correction term $= a + bR + cR^2$
$$= -\,0.272 - 0.140 - 0.220 = -\,0.632$$

The true mean is therefore 0.6 C° or 1.1 F° lower than the arithmetic mean of the extremes at Blue Hill, Mass., in the July average.

Bigelow determined the difference in question empirically and got -0.5 F°. Thus the agreement is not very good as far as the example at random is concerned.

    [7] C E. P. Brooks, "The Reduction of Temperature Observations to Mean of 24 Hours, and the Elucidation of the Diurnal Variation in the Continent of Africa," *Quarterly Journal of the Royal Meteorological Society*, London, vol. 43 (1917), pp. 375–387.

The corrections derived from the formula mentioned above are tabulated in Table 24 for different altitudes between 0 km and 3.5 km and for different average aperiodical ranges between 4°C and 20°C.

TABLE 24   CORRECTION REQUIRED TO REDUCE THE MEAN OF THE EXTREMES (M,m). DIFFERENT AVERAGE NON-PERIODIC DAILY RANGES R (C°) $= M - m$ AND DIFFERENT ELEVATIONS (h, METERS) ARE THE ENTRIES. (AFTER C E P BROOKS)

R(°C) \ h(m)	0	500	1000	1500	2000	2500	3000	3500
4	−0 35	−0 42	−0 49	−0 56	−0 63	−0 70	−0.77	−0 84
6	−0 42	−0 50	−0 58	−0 66	−0 74	−0 82	−0 90	−0 98
8	−0 52	−0 59	−0 65	−0 72	−0 78	−0 85	−0 91	−0 98
10	−0 61	−0 67	−0 70	−0 73	−0.76	−0 79	−0.82	−0 85
12	−0 79	−0 76	−0 73	−0 70	−0 67	−0 64	−0 61	−0.58
14	−0 97	−0 85	−0 73	−0 61	−0.49	−0.37	−0 25	−0 13
16	−1.17	−0 94	−0 71	−0 48	−0 25	−0 02	+0 21	+0 44
18	−1 40	−1 04	−0 68	−0 32	+0 04	+0 40	+0 76	+1 13
20	−1 66	−1 14	−0 62	−0.10	+0 42	+0.94	+1 46	+1.98

Frank H. Bigelow made another step in his investigation.[8] When he had computed the annual course of the *correction* of $(M + m)/2$ to obtain the true mean, he represented the monthly values cartographically and gave isolines of equal corrections for the United States. The result is the exact and concise summary of an immense numerical investigation which is so instructive that at least the two maps of January and July may be offered here. (Figs. 19 and 20.)

The maps show that the trend of the average daily course of temperature is variable from region to region over the continent, as do also the corrections, which give the true mean. It must be emphasized that these variations are not so insignificant. In January they reach 1 F° and in July, 2 F°.

For further purposes, it is important to remember this fact: Mean temperatures derived from extremes without corrections can show differences up to 2 Fahrenheit degrees at places in the United States, even if the true means for an individual month were identical.

Some observatories and meteorological stations attempt to make more than three observations between dawn and dusk in order to get the closest approximation to the true mean. From a

[8] United States Department of Agriculture, Weather Bureau, *Bulletin S* (Washington, 1909).

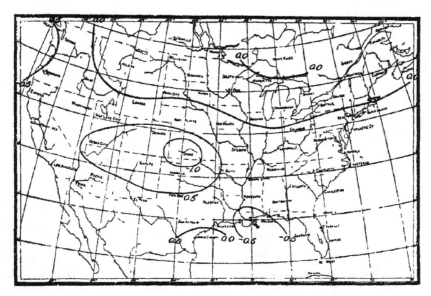

FIG. 19. Correction (F°) to give the "true mean" if $(M + m)/2$ is known.
Isolines for the United States, January.
(After F. H. Bigelow)

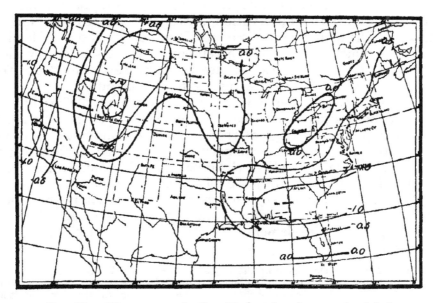

FIG 20. The same as in Fig 19, but for the month of July

first glance at the curve of Figure 18 it is clear that rather the contrary is achieved, because the hours above the daily mean get more and more preponderance.

If the hours of observation are changed in the course of time, or if the daily extreme temperatures are used instead of fixed hours of observations, the corrections vary in a discontinuous manner at the dates of changing. This can be a serious source of error, for instance in calculating periodicities.

The corrections to the plain arithmetic mean vary up to a maximum of about 3/2 F° in the United States, in July, if the hours 7 A.M., 3 P.M., 11 P.M. are used, and reach about −2 F° in the case of the arithmetic mean of the extremes. Therefore, the climatologist who does not know whether the arithmetic means are weighted, corrected, etc., cannot compare the temperatures at different places with one another; or, if he does so, he should know that the accuracy of the comparison is less than perhaps ±2 F°. This reason alone makes it uneconomic to aggravate climatological computations, tables, and graphs, with decimals of degrees Fahrenheit. One exception may be a daily or annual course of temperature derived from some years of observations.

To avoid misunderstandings, let us repeat that all that has been told about the different corrections of the arithmetic means of temperatures at fixed hours of observations, etc., is valid only for averages of at least ten days. Generally these conclusions are applied only to the observation series of a month and longer periods. The average temperature of an individual day is defined as the arithmetic mean of the temperatures at the observation hours, or of the extremes. These averages can differ from the 24 hours' mean of the day by a considerable amount.

The foregoing discussion on the reduction of the arithmetic mean of the records at fixed hours of observation to the true mean is related to the readings of the dry-bulb thermometer. As far as the other climatological elements are concerned, such as vapor pressure, relative humidity, cloudiness, wind velocity, etc., it is usual to identify the arithmetic mean with the true mean.

This is the common practice, but it does not claim that there would not be a better approximation to the true mean. For the greater part of the elements, investigations are lacking in this direction.[9]

[9] The records of cloudiness at the hours of observation, 7 A.M., 2 PM (1 P.M.), 9 P.M., are examined because of the differences from the true mean. (See V. Conrad, *Met Zeit*, 1928, p. 23.) It can be shown that the corrections for the above hours are not significant and can be neglected.

VI. 2. Dates at Which the Average Temperature Crosses Certain Thresholds

Generally, these dates are taken from graphs (drawn on a large scale) which represent the annual course. As this method is laborious, a simple interpolation is indicated here. The formula is founded on the fact that the trend of the ascending and descending branches of the curve is linear in rough approximation. Therefore, the interpolation method is no longer applicable near the extremes.

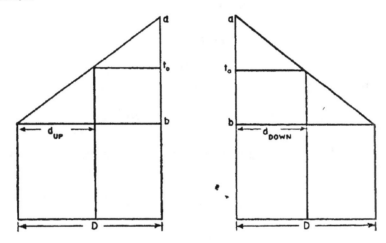

Fig. 21.   Diagram explaining the interpolation of the dates at which temperature crosses a given threshold upward and downward   (After V Conrad)

The symbol t_o means the given threshold of temperature, a the monthly average above it, and b that next below it. The letter D is the difference in days between the middles of the months with the average temperature (b or a) and the middle of the consecutive month (see Fig. 21). The symbol d_{up} is the difference in days between the middle of the month below the threshold and the date at which the temperature (t_o) of the threshold occurs.

According to G. Crestani ("Alcuni considerazioni sul calcolo della media della velocità del vento, sull'andamento diurno ed anno della medesima in Italia," *Boll. bimens.* XLIX, 1930, 10–14), the combination of wind velocities, $\frac{1}{3} \cdot$ (9 A.M. + 3 P.M + 9 P.M.), yields corrections to the true mean which are zero in February and November and reach their maximum in May with about −7% of the true mean   On the average for the year, the correction is −4% of the true mean.

If the velocity of the wind is only estimated, not instrumentally measured, the arithmetic mean of the above-mentioned observation hours shows negligible corrections to the true mean. Probably the same is valid with three other reasonable hours of observation.

The symbol d_{do} is the same for the descending branch of the curve. Then:

$$d_{up}:D = (t_o - b).(a - b)$$

and

$$d_{do}:D = (a - t_o):(a - b)$$

Hence

$$d_{up} = D\frac{t_o - b}{a - b}$$

$$d_{do} = D\frac{a - t_o}{a - b}$$

It is sufficiently accurate to assume

$$D = 30 \text{ days}$$

Finally the interpolation formula reads:

$$d_{up} = 30\frac{t_o - b}{a - b} \qquad d_{do} = 30\frac{a - t_o}{a - b}$$

As an example, the annual course of temperature at Bismarck, N. D. (Table 25), will serve.

TABLE 25   THE ANNUAL COURSE OF TEMPERATURE AT BISMARCK, N D
(46 8°N, 100 6°W, 1670 FEET)

		Deviations		
	°F (*)	°F	°F–Jan	Rel Temp %
Jan	8 6	−32 0	0 0	0
Feb.	9 9	−30 7	1.3	2
Mar	24 3	−16 3	15 7	26
Apr	43 3	+ 2 7	34 7	56
May	54 3	+13.7	45 7	74
June	64 2	+23 6	55 6	91
July	70 0	+29 4	61 4	100
Aug	67 5	+26 9	58 9	96
Sept.	57 9	+17.3	49 3	80
Oct	44 6	+ 4 0	36 0	59
Nov.	28 0	−12 6	19 4	32
Dec.	14 7	−25 9	6 1	10
Year	40.6			

* After R DeCourcy Ward and C F. Brooks, "The Climates of North America," Köppen-Geiger, *Handbuch der Klimatologie*, pt. J (Berlin, 1936).

The dates when the temperature rises above and drops below 32°, 43°, 65° should be calculated.   The threshold of 32° divides the curve into the parts below and above the freezing point; 43° is about the temperature limit of germination of seeds, 65° is the assumed temperature limit for artificial heating.

$$d_{up} = 30 \, \frac{32 - 24.3}{43.3 - 24.3} = 12$$

and

$$d_{do} = 30 \, \frac{44.6 - 32}{44.6 - 28.0} = 23$$

$d_{up} = 12$ is to be added to March 16 and
$d_{do} = 23$ to October 16

Thus we see that:

> The temperature rises above 32°: March 28
> The temperature drops below 32°: November 8

In the same way it follows that:

> The temperature rises above 43°: April 15
> The temperature rises above 65°: June 19
> The temperature drops below 65°: August 24
> The temperature drops below 43°: October 19

VI. 3. DURATION OF TEMPERATURES ABOVE AND BELOW CERTAIN THRESHOLDS

These data are very characteristic for the annual course of temperature and should not be omitted.

If a table is at hand with running numbers of the dates from January first, the calculation is nothing but a simple subtraction.[10] The duration in days for the three thresholds mentioned, at Bismarck, N. D. (see VI, 2), is:

> above 32°F: 225 days = 32 weeks = 62% of the year
> above 43°F: 187 days = 27 weeks = 51% of the year
> above 65°F: 66 days = 9 weeks = 18% of the year

Sometimes the conversion of the number of days into weeks can be recommended, if the observations are not very accurate, as also when a quick and easy survey is given. This purpose can be better attained with the expression that yields the smaller numbers. Often also the knowledge of the percentage of the year with temperatures above and below the thresholds is useful.

[10] See Annex IV.

It is instructive to learn that at Bismarck, N. D., the average temperature remains below the freezing point for 38% of the year and above for 62%, or that only 18% of the year has a temperature above the technically assumed threshold for artificial heating.

The duration of a temperature higher than 43°F is often used in climatological investigations and is called the *vegetative period*. In the United States the term *growing season* is more or less identical with "vegetative period." It means the space of time between the last and the first *killing frost*, or, in the absence of satisfactory frost data, between the last and first minima of 32°F or lower.

Because I could not find a proper definition of "killing frost" in the available literature, Dr. C. F. Brooks was kind enough to give me the following information. I quote: "A killing frost is one which kills the general vegetation at a place. There is no specific meteorological definition. Where the vegetation is succulent and easily damaged by frost, a killing frost will occur at a higher temperature than where the vegetation is hardy. In the absence of killable vegetation, a temperature of 32°F in a thermometer shelter is taken as a killing frost. Generally, however, the killing frost, first in fall and last in spring, occurs with a minimum shelter temperature higher than 32°, sometimes as high as 40°F."

Therefore, "killing frost" and "growing season" are generally defined biologically, not quantitatively.[11]

It goes without saying that the principle of duration is applicable also to many other climatological elements, as average duration of a certain wind direction, of the rainy or hot period of the year (see Figs. 40–41), of the snow-cover, of any average state of the atmosphere or of the surface of the ground, as far as periodical climatological phenomena are concerned. In all these cases, any information about the respective durations is a good contribution to the knowledge of the phenomenon.

VI. 4. OTHER CHARACTERISTICS OF TEMPERATURE CONDITIONS. SPELLS OF COLD AND HOT DAYS

Certain thresholds and definitions must be mentioned. In order to discriminate between (a) *ice days* and (b) *frost days*, "ice days" are defined as the days with the maximum below the freezing point, while "frost days" have the minimum below the freezing point.

[11] See also W. G Reed, "Frost and the Growing Season," in *Atlas of American Agriculture* (U S. Dept. Agr., Washington, D C , 1918).

Days with a maximum ≧ 77°F are called (c) *summer days*, and days with a maximum ≧ 86°F are called (d) *tropical days*.

All these thresholds are arbitrary. But they are used frequently and should be chosen in any case; others can be added which appear advantageous for special purposes. For such thresholds, frequencies should be computed. On the other hand, the thresholds are used for calculating spells of frost days, ice days, summer days, tropical days, etc.

If some years' observations are at hand, the lengths of the respective spells are evaluated for each January, February, etc., of the series. Then, the average length and the greatest length of the spells should be given.

If we have to handle ice days, summer days, tropical days, the temperature-maximum of the consecutive days is decisive for the length of the spells. In the case of frost days, it is the minimum that is decisive.

In other investigations, the daily mean of consecutive days is considered. As far as the technique of counting out the duration of spells is concerned, the following rule is valid: If the spell begins in one month and ends in the next, the whole length in days is added to that month to which the greater part belongs; if the spell is longer than two consecutive months, the two months have to be joined.

VI. 5. Cumulated Temperatures

Starting from a certain threshold, the mean daily temperatures above or below it are added together for the single month, for the season, or for the year. If some years' observations are available, the respective sums are averaged for the whole period.

Especially in the earlier agricultural, meteorological, and climatological investigations, the mean daily temperatures above 43°F (the temperature of germination as stated earlier) are totaled for the single month or for the growing season, etc.

A. Angot calculated totals of daily minima at or below the freezing point. Table 26 gives these sums for the region of western

TABLE 26. Average Cumulated Daily Minimum Temperatures ≦ 0°C for Paris, Winter 1872/73 to 1911/12 (After A. Angot)

				Centigrade degrees				
Oct	Nov	Dec	Jan	Feb	Mar	Apr	May	Year
3.3	16.1	51.7	59.8	42 7	21.9	3.0	0 2	198.7

Europe which is characterized by a climate of slight continentality. This procedure permits also a quantitative distinction between severe and mild winters, if a classification is made according to the standard deviation of the annual sums of minimum temperatures $\geqq 0°C$. (See III, 3.)[12]

VI. 6. DEGREE DAYS

The threshold for artificial heating is a daily mean of 65°F, according to the experience of heating engineers. If the daily mean drops below this temperature, heating is necessary and the amount of fuel to maintain a comfortable inside temperature is more or less intimately correlated with the number of degrees below 65°F. This difference (65°F minus each day's mean) is called number of *degree days*. The degree days are totaled for the month, the season, and the year, counting negative differences as zero. If a series of some years' observation is available, the cumulated degree days of the month, etc., are averaged.

TABLE 27. DEGREE DAYS (F°) AT BOSTON, MASS. (42 4°N, 71 1°W, 125 FT) AND CHICAGO, ILL (41 9°N, 87 6°E, 673 FT)
(Boston 1914/15 to 1942/43, Chicago 1871 to 1936)

	Boston	Chicago
Jan.	1101	1256
Feb	1027	1083
Mar	852	903
Apr.	538	545
May	248	280
June	66	69
July	8	8
Aug	16	11
Sept	98	94
Oct	338	363
Nov.	651	742
Dec.	1000	1112
Year	5943	6466

Table 27 gives the cumulated degree days for Boston, Mass.,[13] and Chicago, Ill [14] The numbers for Boston are derived from 28

[12] See V Conrad, "Die klimatologischen Elemente und ihre Abhängigkeit von terrestrischen Einflüssen, Koppen-Geiger, *Handbuch der Klimatologie*, vol. I B (Berlin, 1936), p. 112.

[13] I have to thank Mr Edward Sable of the U S Weather Bureau for the Boston data

[14] Taken from Thomas A. Blair, *Climatology* (New York, 1942), p. 18.

heating seasons (1914/15 to 1942/43), those for Chicago from the period 1871 to 1936. The comparison between the two series may be interesting.[15]

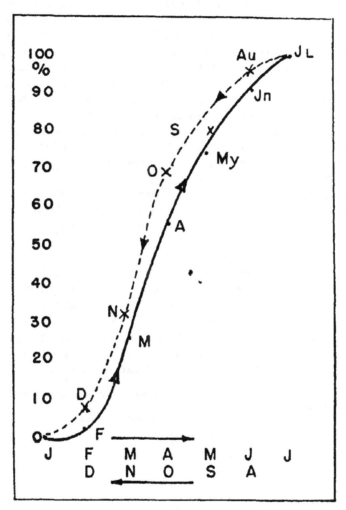

FIG. 22. Relative temperatures at Bismarck, N. D.

VI. 7. RELATIVE TEMPERATURES

The comparison of the annual course of temperature at one place with that at another place presents two difficulties:

[15] In spite of unequal periods.

1) The average temperatures of the year are different.
2) The amplitudes (annual ranges) are different.[16]

The first inequality is overcome by calculating deviations from the arithmetic mean (annual average). (See Table 25, deviations.) The first and second are eliminated by the method of *relative temperatures*, a concept introduced by W. Koppen. The degrees by which the consecutive monthly averages exceed the temperature of the coldest month are each expressed in per cent of the difference between the average temperatures of the warmest and coldest months. By this explanation, the calculation of the numbers of the last column in Table 25 is easily understood.

The author has given a graphical representation of relative temperatures [17] which facilitates the comparison of different series of relative temperatures. These are plotted at equal intervals to the right for the first seven months (Fig. 22).

Starting from July, the curve turns back to the left, showing the values of the other months. Every point bears the initial of the month. Furthermore, the curve of relative temperatures for January through July is a full line, while the curve of the other half of the year is represented by a broken line.

The asymmetry of the trend of the annual course of temperature at Bismarck, N. D., is evident. The months after July are warmer than the symmetrical ones before July. August is warmer than June, September warmer than May, and so forth. In this and in other respects, the representation in Figure 22 gives a good analysis of the annual curve, free from the embarrassing influence of the annual average and of the size of annual variation.

[16] The reader should distinguish between annual range and amplitude. The latter is half of the periodical range
[17] V. Conrad, *Fundamentals of Physical Climatology*, p 28.

CHAPTER VII

ATMOSPHERIC PRESSURE AND PRESSURE
OF WATER VAPOR

PRESSURES are statistically discussed in the same way as temperature. It should be emphasized that water-vapor pressure is a unilaterally limited element, and there are vast regions where the vapor pressure is practically zero in winter. It is not the aim of this book to deal with the different physical definitions regarding the water-vapor content of the atmosphere. For this, see V. Conrad, *Fundamentals of Physical Climatology*, and for a more detailed treatment, Sir Napier Shaw, *Manual of Meteorology*.

CHAPTER VIII

WIND

VIII. 1. RESULTANT WIND VELOCITY CALCULATED FROM RECORDS OF DIRECTION AND VELOCITY

THE RECORDS of an anemograph yield an average direction of the wind for each hour, and the run of the wind, in respective units.

The example, Table 28, shows the statistics derived from continuous wind records at Batavia, Java, in January 1926. The first line F gives the number of hours with the various directions

TABLE 28   EXAMPLE OF THE CALCULATION OF THE WIND-ELEMENTS; BATAVIA JAVA (6 2°S, 106 8°E, 26 FT), JANUARY 1926*

		N	NE	E	SE	S	SW	W	NW
I	Number of hours (F)	153	37	1	8	41	25	92	137
II	Run of wind (miles) (L)	309	56	(1)	4	29	29	176	465
III	Average velocity ($v = L/F$) (mi/h)	2.01	1 52	(1.34)	0 56	0.72	1.16	1.92	3.40

* Small differences of v against the ratios L/F originate from the conversion of meters per second (m/s) into mi/h.

during the period in question. The second line shows the run of the wind L for each direction during the same period (here the month of January). In the third line, the ratio L/F appears, i.e., the average velocities of the respective directions. The *resultant* run of the wind is calculated by Lambert's formula [1]

$$C_N = N - S + (NE + NW - SW - SE) \cos 45°$$
$$C_W = W - E + (NW + SW - NE - SE) \cos 45°$$

where C_N and C_W are the total run of the wind from north to south (subscript N) and from west to east (subscript W). If places only in temperate latitudes of the northern hemisphere are considered, it is recommended that the positive sign be given to winds from the north and from the west. It is practical and sufficiently exact

[1] Auxiliary tables for the use of this formula are given in the Smithsonian Tables.

to make the computation with the value cos 45° = 0.7, so that the calculation can be worked out mentally.

$$C_N = N - S + 0.7 (NE + NW - SE - SW)$$
$$C_W = W - S + 0.7 (NW + SW - NE - SE)$$

For 16 directions we get:

$$C_N = N - S + (NNE + NNW - SSE - SSW) \cos 22\tfrac{1}{2}°$$
$$+ (NE + NW - SE - SW) \cos 45°$$
$$+ (ENE + WNW - ESE - WSW) \cos 67\tfrac{1}{2}°$$

$$C_W = W - E + (WNW + WSW - ENE - ESE) \cos 22\tfrac{1}{2}°$$
$$+ (NW + SW - NE - SE) \cos 45°$$
$$+ (NNW + SSW - NNE - SSE) \cos 67\tfrac{1}{2}°$$

(The approximations:
$$\cos 22\tfrac{1}{2}° = 0.9$$
$$\cos 45° = 0.7$$
$$\cos 67\tfrac{1}{2}° = 0.4 \qquad \text{are mostly sufficient}).$$

Generally, one can be content with 8 directions. If 16 directions are given in the original data, it is usual to reduce these to 8 directions. The frequencies of the intermediate directions are allotted by 50% to the two nearest main directions.[2] Analogous formulas can be given for the points of the compass.[3]

The values L of the numerical example of Table 28 [4] calculated with Lambert's formula yield:

$$C_N = + 622 \text{ miles for January 1926}$$
$$C_W = + 479 \text{ miles for January 1926}$$

Therefore

$$C_N/C_W = \tan a' = + 622 / + 479 = 1.30$$

or $a' = 52°$, the angle between the north and the west direction

[2] Objections to this procedure are more or less justified theoretically, but in practice this method is a sufficient approximation.

[3] If a greater accuracy is desired or if more than 8 directions have to be considered, L. W. Pollak's *Rechentafeln zur harmonischen Analyse* (Leipzig, 1926) facilitate the procedure. The "Handweiser zur harmonischen Analyse," *Prager Geophysikalische Studien*, vol. II (Prague, 1928), pp. 54–59, contains special instructions for calculating the resultant wind direction.

[4] Data taken from *Observations Made at the Royal Magnetical and Meteorological Observatory at Batavia* (Batavia, 1930), with some handwritten corrections by H. P. Berlage, Jr.

(Fig. 23). Because the azimuth of wind direction is usually reckoned in a clockwise direction (north through east, etc.), the true azimuth

$$a = 52° + 270° = 322°$$

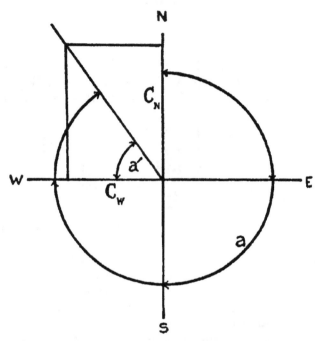

FIG. 23. Graph for evaluating the true azimuth of the resultant wind

VIII. 2. RESULTANT WIND DIRECTION CALCULATED FROM FREQUENCIES OF THE DIRECTIONS ALONE

It often happens that only the frequencies of the single wind directions are published, while their average velocities are not known.   In this case, also, it is possible to estimate the resulting wind direction.

The method is based on the assumption that the average velocities are roughly proportional to the frequencies of the directions.[5]   The author examined the wind records for 35 months at Batavia, Java.   The correlation between frequency of the directions and the velocity is shown in Figure 24.   If, as in this random sample, the assumption in question holds, then winds from the

[5] For instance, if west winds are the most frequent winds at a place, they are also the strongest winds.

most frequent directions are also the strongest. Hence, in the first approximation, the frequencies F (first line of Table 28) can serve for L, the run of the wind. If we calculate with the values F of Table 28, we get

$$a = 321°$$

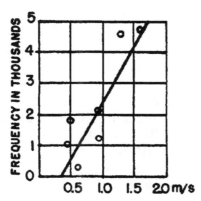

FIG 24. Correlation between the frequency of the directions and the average velocity of the wind, demonstrated by means of the wind records of 35 January months at Batavia, Java. (After V. Conrad)

instead of the true value, $a = 322°$. In this example the difference is incidentally negligible.[6]

VIII. 3. The Resultant Run of the Wind

The resultant run of the wind is given by

$$\Re = \sqrt{C_N{}^2 + C_W{}^2} = \sqrt{622^2 + 479^2} = 785 \text{ miles}$$

for Batavia, Java, January 1926.

[6] The calculation which leads to the result given above is as follows:
The numbers of hours with the different wind directions (F) (Table 28, first line) are now introduced into Lambert's formulas instead of the values L.

$$C_N = (153 - 41) + 0.7(37 + 137 - 8 - 25) + = 210\,7$$

$$C_W = (92 - 1) + 0\,7(137 + 25 - 37 - 8) = + 172.9$$

Therefore

$$\tan a' = + 210.7/+ 172\,9 = 1\,2186$$

or

$$a' = 50° 38'$$

and finally

$$a = 320° 38'$$

The average resultant wind velocity $\bar{v}$ is calculated by dividing $\Re$ by the number of hours with wind. That means that the number of calms is excluded. The sum of the hours with measurable wind velocity is in the example of Table 28 (first line) $\Sigma F = 494$ Therefore

$$\bar{v} = \frac{\Re}{F} = \frac{785}{494} = 1.59 \text{ mi/h} = 0.71 \text{ m/s}\,[7]$$

Frequently, Beaufort numbers are published in the year-books and annals; these should be converted into absolute measure, i.e., m/s or miles per hour, etc., for purposes of comparison. Therefore, Table 29 gives equivalents in meters per second, kilometers per hour, and miles per hour. These equivalents were recommended by the meeting of the International Meteorological Committee at Vienna, Austria, in 1927.[8]

TABLE 29   Equivalents of the Numbers of Beaufort Scale *

Beaufort numbers	m/s	km/hour	miles/hour
0	0 25	0 9	0.6
1	1 15	4	2.6
2	2 55	9	6
3	4 30	16	10
4	6 40	23	14
5	8 65	31	19
6	11 15	40	25
7	13 85	50	31
8	16 75	60	38
9	19 90	72	44
10	23.35	84	52
11	27.10	98	61
12	>29	>104	>65

* The average values are calculated from the limits in m/s, originally given by the International Meteorological Committee (1927) for each Beaufort degree.

In Table 29, the Beaufort no. 0 is equivalent to 0.25 m/s, owing to the fact that the record "0" (calm) includes also the very weak winds.

[7] The necessary tables for conversion of the English measure to the cgs-system, respectively centigrades, meters, etc, are given in the Smithsonian Meteorological Tables (Washington, 1918).

[8] See also The Meteorological Glossary (Meteorological Office, London, 1939), p. 36, with slightly different values.

VIII. 4. The Steadiness of the Wind (S)

The steadiness of the wind is the ratio of the resultant run of the wind ($\Re$) to the run of the wind (R), disregarding the direction, multiplied by 100.

$$S = 100 \frac{\Re}{R}$$

If during the whole period considered, the wind always blew in one and the same direction, then

$$\Re = R$$
$$S_{Max} = 100\%$$

and we have the upper limit of S. If there is no prevailing wind direction and the wind shifts in a quite incidental way and there are a great many wind observations, the sum of all velocities will be zero if their direction is considered. Then $\Re = 0$, while R has a finite value (because a place with everlasting calm is impossible). Therefore the lower limit

$$S_{Min} = 0\%$$

The steadiness of the wind varies between 0 and 100%, according to the above definition.

In the example of Table 28, the total run of the wind, disregarding the direction (sum of second line)

$$R = 1069 \text{ miles}$$

The resultant run of the wind (considering the direction) is:

$$\Re = 785 \text{ miles}$$

Therefore

$$S = 100 \frac{785}{1069} = 73\%$$

The complete result may be expressed as follows: The rainy monsoon at Batavia, in January 1926, had an average net velocity of 1.6 miles per hour, an average azimuth of 322° (W 52°N) (N 38°W) and a steadiness of 73%. These indications characterize the average wind conditions. The great steadiness of the monsoon wind in January is very significant for this phenomenon.

CHAPTER IX

SOME COMBINED ELEMENTS

IX. 1. RELATIVE HUMIDITY (R.H.)

A GRAPHICAL correlation between the monthly average relative humidity and the average temperature, called *climogram*, is frequently used. Figure 25 gives an example for Boston, Mass.[1] The method may be valuable for purposes of illustration. In the present example, it might be stated that the

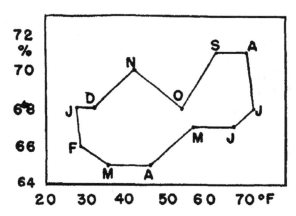

FIG. 25. Climogram of Boston, Mass. (42 4°N, 71.1°W, 125 feet)

months August through December are more humid than the months March through June, at similar temperatures on the New England coast.

This kind of representation is, of course, not restricted to relative humidity and temperature. It can be used for every pair of elements which are subject to an annual course and are related to one another. Fine examples are presented by P. Götz,[2] who correlates in this way, for instance, the red and infrared content of radiation, turbidity factor, etc., with water-vapor pressure.

[1] G H. Noyes, *Annual Meteorological Summary with Comparative Data 1942* (U. S. Weather Bureau, Boston, 1943) See also F Eredia, "Sul' umidità relativa in Italia," Gerland's *Beiträge zur Geophysik*, vol. 33 (1931), p 286.
[2] F W Paul Götz, *Strahlungsklima von Arosa* (Berlin, 1926).

IX. 2. Equivalent Temperature (Θ)

Equivalent temperature combines temperature with water-vapor pressure. For illustration: the total water-vapor content of one cubic meter could be condensed at constant pressure. By this process a certain amount of heat would be released, with which the dry air could be heated Δt C°.[3]

If t is the observed air temperature, then the equivalent temperature is defined by the equation:

$$\Theta = t + \Delta t$$

For usual climatic temperatures, the difference

$$\Delta t = ke$$

which means, it is proportional to the water-vapor pressure, e, where k is the coefficient of proportionality. This coefficient is expressed by the following equation, given by F. Linke:

$$k = \frac{R(607 + 0.708t)(273 + t)}{0.239 \times 13\ 6 \times b \times 1000}$$

where R is the gas constant ($R = 29.27\ g_{45}m^2\ \sec^{-2}$), t = temperature in degrees centigrade and b means the air pressure in millimeters of mercury (Hg). This formula has been evaluated for different pressures and temperatures between $b = 770$ mm Hg and $b = 740$ mm, within the limits of the temperatures $-20°C$ $(-4°F)$ and $30°C$ $(86°F)$. The calculation yields the following results. if

$t =$	-20	and	$b = 770$ mm:	$k =$	1.84
$t =$	$+30$	and	$b = 740$ mm:	$k =$	2.16

In the case of simple estimations where no real exactness is required, and at medium pressures and temperatures, it can be assumed that $k \doteq 2$. Thus

$$\Theta = t + 2e$$

---

[3] The total heat content (W) of a cubic meter of moist air is composed of the heat contents of its dry air and its water vapor. The equation reads:

$$W = c_p \rho T + rf \text{ kgcal,}$$

where c_p means the specific heat of air at constant pressure, ρ the density, T the absolute temperature, f the absolute humidity in grams per cubic meter, and r the heat of vaporization. Dividing the total heat content by $c_p \rho$, we obtain a value of temperature character.

In rough approximation, the absolute humidity, f (g per m³) and the vapor pressure e are more or less equal numbers as far as usual climatic temperatures are concerned (See F. Linke, *Met. Zeit.*, 1922, pp. 268 ff.)

This formula is valid as an approximation only within the limits mentioned and should not be used if these limits are exceeded by temperature and pressure. As an example, the annual course of the average equivalent temperatures at Boston, Mass., is given (Table 30, columns 5 and 6). The other columns present the numerical values for calculating the equivalent temperature.

Finally, in some cases it is desirable to figure out *exact* values of the equivalent temperature. F. Linke[4] has given another equation:

$$\Theta = t + RH \cdot K(t, b)$$

where

$$K(t, b) = \frac{1543 + 1.68t}{b - 0.377E} \frac{E}{b}$$

and $RH = e/E$.

For evaluating the formula, the temperature t in centigrade degrees, the air pressure in millimeters of mercury, and the relative humidity have to be known. The table in Appendix III is a good expedient, for this computation and yields directly the quantity of $K(t, b)$. The vertical entry is the temperature, the horizontal, the air pressure. Examples follow:

1) $t = -18°C;$ $b = 780$ mm; $RH = 0.70$
    from the table: $K = 1.87$
    $\Theta = -18 + 0.70 \times 1.87 = -16.7°C$

2) $t = 0°C;$ $b = 760$ mm; $RH = 0.70$
    from the table: $K = 9.32$
    $\Theta = 0 + 0.7 \times 9.32 = 6.5°C$

3) $t = 30°C;$ $b = 740$ mm; $RH = 0.70$
    from the table: $K = 65.2$
    $\Theta = 30 + 0.7 \times 65.2 = 75.6°C$

The examples show the simple procedure of calculation by means of the table and the interesting effectiveness of vapor content upon the equivalent temperature at different air temperatures.

The calculation is still simpler if the temperature of the wet bulb t' is known. Then

$$\Theta = t' + K(t', 760)$$

[4] F. Linke, *Meteorologisches Taschenbuch* (Leipzig, 1931), p 287.

Example: if

$$t' = 25.9°C, \qquad \Theta = 75.7°C$$

This wet-bulb temperature corresponds to the temperature-humidity conditions of example no. 3. The two calculations of this example yield identical results. In the last equation, correlating t' with Θ, one needs consider neither the actual air pressure nor the actual relative humidity.

TABLE 30. AVERAGE MONTHLY EQUIVALENT TEMPERATURES
($\Theta = t + 2e$) AT BOSTON, MASS.

	°F (1)	°C (2)	e, mm (3)	2e (4)	°C (5)	°F (6)
Jan..	27.9	−2.3	2.6	5.2	2.9	37
Feb..	28 8	−1.8	2 6	5.2	3.4	38
Mar. ..	35 6	2 0	3 4	6.8	8.8	48
Apr. .	46.4	8.0	5 2	10 4	18.4	65
May	57.1	13.9	7 9	15 8	29.7	85
June . .	66 5	19.2	11.2	22 4	41.6	107
July. ..	.71.7	22.1	13.6	27.2	49.3	121
Aug	69 9	21.1	13.3	26 6	47 7	118
Sept ...	63 2	17.3	10 5	21.0	38.3	101
Oct..	53 6	12 0	7 2	14 4	26 4	80
Nov..	42 0	5 6	4 8	9 6	15 2	59
Dec. . .	32.5	0 3	3.2	6 4	6 7	44

Column 1: Air-temperature in °F.
Column 2: Air-temperature in °C.
Column 3: Water-vapor pressure in millimeters of mercury.
Column 4: The values of Col. 3 doubled
Column 5: Equivalent temperature in °C
Column 6: Equivalent temperature in °F

TABLE 30a. THE EQUIVALENT TEMPERATURES COMPUTED FROM $\theta = t + 2e$ (TABLE 30), COMPARED WITH THE RESULTS (Θ') FROM THE CORRECT FORMULA (SEE P. 109).*

	°C	RH	k(t, 760)	K × RH	Θ'°C	Θ°C	(Θ' − Θ) °C
I	−2.3	0.67	7 83	5.2	2 9	2.9	0 0
II	−1.8	.64	8 13	5 2	3 4	3.4	0.0
III	2 0	.65	10 7	7 0	9 0	8.8	+0 2
IV	8 0	.65	16 2	10 5	18 5	18.4	+0.1
V	13.9	.66	24 0	15 8	29.7	29.7	0.0
VI	19 2	.67	33 4	22 4	41 6	41 6	0 0
VII	22.1	.68	39.7	27.0	49 1	49 3	−0.2
VIII	21.1	.71	37.4	26.6	47.7	47.7	−0 0
IX	17.3	.72	29.6	21.3	38 6	38.3	+0.3
X	12.0	67	21 1	14.1	26.1	26 4	−0.3
XI	5.6	.70	13.8	9.7	15.3	15.2	+0.1
XII	0.3	.67	9.53	6 4	6 7	6.7	0.0

* $K(t,b)$ is taken from Appendix III, assuming $b = 760$ mm Hg, constant through the year.

IX. 3. Drying Power

Drying power combines saturation pressure for a given actual or average temperature, water-vapor pressure, and wind velocity. A short instruction on how to calculate this combined element [5] may suffice. The theoretical basis could be improved. The idea is extremely useful in the description of a climate. The definition reads: *Drying Power is the amount of evaporation, in centimeters, within four hours.* It is calculated by means of the following empirical formula:

$$S_v = 0.023 \times F(v) \frac{E_t}{e} \times \frac{\Delta E}{\Delta t} (1 + 0.084 \, v)$$

where

S_v = drying power at the wind velocity, v, in km/hour
e = the observed water-vapor pressure in mm Hg
E_t = saturation pressure at the temperature, $t°$C
$F(v)$ = a factor dependent on v [6]

$\Delta E/\Delta t$ means the variation of the saturation pressure with temperature and can be taken from any psychrometric table which indicates temperature in °C and the saturation pressure in mm Hg.

Example: The saturation pressure, E, is taken at one degree above and at one degree below the given air temperature. If $t = 10°$C then from the table $E_{9°} = 8.61$ mm and

$$E_{11°} = 9.84 \text{ mm} \qquad \text{then} \qquad \frac{\Delta E}{\Delta t} = \frac{9.84 - 8.61}{11 - 9} = 0.615$$

If $\Delta E/\Delta t$ is known, the computation of the formula of the drying power offers no further difficulty or doubt.

[5] W. Knoche, "El 'Valor de Desecación' como factor climatológico," *Revista Chilena de Historia y Geografía* 1919, nos. 34, 35; W. Knoche, "Der 'Austrocknungswert' [Drying Power] als klimatischer Faktor," *Aus dem Archiv der Deutschen Seewarte*, vol 48 (1929), no. 1.
[6] The values of $F(v)$ for different wind velocities are:

km/hour	$F(v)$	km/hour	$F(v)$
0	1.000	6	1.667
1	1.148	7	1.712
2	1 274	8	1 746
3	1.392	9	1.762
4	1.493	10	1.777
5	1.592	15	1 782

It is only necessary to add that the formula is valid only for sea-level. For other levels, S_v has to be multiplied by b_o/b, where $b_o = 1015$ mb and b is the pressure in millibars at the place in question.

IX. 4. Cooling Power

Cooling power is observed by the dry and by the wet kata-thermometer invented by Leonard Hill. This book does not deal with instrumental technique, but it may be noted that Hill succeeded so far in analyzing the cooling components as to arrive at analytical equations which give the cooling power from values of temperature, wind velocity, and humidity.

Cooling power is a bioclimatological combined element and is defined as follows: *The cooling power is equal to the heat loss which a surface of a body at the temperature of the human blood (99°F, 37°C) experiences, if exposed to the free air.* It is measured in milligram calories per square centimeter and second (mgcal/cm², sec).

There are two kinds of cooling power:

1) Heat loss of a dry surface (H).
2) Heat loss of a wet surface (H').

The equations read:

A) for wind velocities less than 1 m/s:

dry surface: $H = (0.20 + 0.40\sqrt{v})(36.5 - t)$
wet surface: $H' = H + (0.060 + 0.072v^{1/3})(E - e)^{4/3}$

  H = the dry cooling power in mgcal/cm², sec
  H' = the wet cooling power in mgcal/cm², sec
  v = the wind velocity in meters per second
  t = the air tempersture in °C
  E = the saturation pressure in millimeters of mercury [7]
  e = the vapor pressure in millimeters of mercury

B) for wind greater than 1 m/s:

dry surface: $H = (.13 + .47\sqrt{v})(36.5 - t)$
wet surface: $H' = H + (.035 + .098v^{1/3})(E - e)^{4/3}$

In English units, i.e., when v is the velocity in feet per minute and t the temperature in degrees Fahrenheit, the equations read:

[7] At 36 5°C (97 7°F).

dry surface:

below 200 ft/min $H = (.11111 + .01584\sqrt{v})(97.7 - t)$

above 200 ft/min $H = (.07222 + .01861\sqrt{v})(97.7 - t)$

wet surface·

below 200 ft/min $H' = (.19444 + .08118\sqrt{v})(97.7 - t')$

above 200 ft/min $H' = (.05556 + .10505\sqrt{v})(97.7 - t')$

The formulas for wind velocities greater than 1 m/s or 200 ft/min seem to hold very well; but are perhaps not fully valid for the small velocities.[8]

The symbol t' of the two equations for the wet surface denotes the temperatures of the wet-bulb thermometer.

Newer formulas for the dry cooling power originate from particularly exact experiments.[9]

$$H = 0.99\sigma(T_K{}^4 - T_A{}^4) + (.113 + .34v^{0\ 622})(T_K - T_A)$$

where T_K denotes the absolute temperature of the kata-thermometer, T_A the temperature of the air, and σ the constant of the Stefan-Boltzmann-law:

$$\sigma = 8.26 \times 10^{-11}\ \text{gcal cm}^{-2}\,\text{min}^{-1}\,\text{deg.}^{-4}$$

For moderate conditions, i.e.,

$$(T_K + T_A):2 \doteq 300°a, \quad \text{in very rough approximation,}$$

the formula above can be simplified.[10]

$$H = (t_K - t_A)(.26 + .34v^{0\ 622})$$

where t_K and t_A mean centigrade degrees.

[8] V Conrad, "Messung und Berechnung der Abkuhlungsgrosse," Gerland's *Beitrage zur Geophysik*, XXI (1929), 183 For further information See D Hargood Ash and Leonard Hill, "The Kata-thermometer," in *Studies of Body Heat and Efficiency* (London, 1923), C F Brooks, "The Cooling of Man under Various Weather Conditions," MWR, vol. 53 (1925), pp 423–424, E C Donnelly, "Human Comfort as a Basis for Classifying Weather," *ibid.*, pp 425–426, V. Conrad, *Fundamentals of Physical Climatology*, where the reader will find a diagram (p 113) from which he can make rough estimations of dry cooling power for temperatures between −50°F and +98°F and wind velocities from calm to 25 mi/h

[9] H Lehmann, *Mikroklimatische Untersuchungen der Abkühlungsgrosse in einem Waldgebiet* (Leipzig, 1936). See also R G Stone, "On the Practical Evaluation and Interpretation of the Cooling Power in Bioclimatology," BAMS, vol 24 (October 1943), pp. 304 ff.

[10] K. Büttner, *Physikalische Bioklimatologie* (Leipzig, 1938), p. 98.

For bioclimatological purposes t_x has to be assumed as constant at 36.5°C. It should also be emphasized, from the methodological standpoint, that the relations of cooling power to physiological processes are very complicated. The application of cooling power to the problems of the heat balance [11] of the human body assumes that the temperature of the human skin is constant. This assumption does not hold.

[11] See C F. Brooks and E. C. Donnelly, cited in note 7 above, and D. Brunt, in QJRMS, 1943.

CHAPTER X

CLOUDINESS

IN THE FOREGOING DISCUSSION of variates of different kinds, the statistical properties of cloudiness were considered, and their relation to the relative sunshine duration too. (See III. 6 and III. 10.) Especially for this element, average values have to be supplemented by computing frequency distributions.[1] When observing and discussing cloudiness, one should discriminate between low and high clouds.[2]

X. 1. CLEAR AND CLOUDY DAYS

In a number of national meteorological networks no detailed observations of the cloudy part of the sky are made. Only the day as a whole is characterized "clear," "cloudy," or "partly cloudy." In the United States, this kind of observation is usual at the coöperative stations, because records are made only once a day.

The international definitions (Utrecht, Holland, 1874) are:

1) *Clear* days are days with an average cloudiness between 0% and 20% of the visible sky.

2) *Cloudy* days have an average cloudiness between 80% and 100%.

The instructions for coöperative observers of the U. S. Weather Bureau which differ from the international use are as follows: "The general character of the day from sunrise to sunset should be recorded as 'clear' when the sky averages three tenths or less obscured . . . and 'cloudy' when more than seven tenths obscured."

A successful effort has been made to correlate the frequencies of clear and cloudy days with the average percentage of cloudiness. The records at stations where cloudiness as well as frequencies of clear and cloudy days are simultaneously observed, give the opportunity of establishing an analytical relation between cloudiness

---

[1] W. Köppen, H. Meyer, "Die Häufigkeit der verschiedenen Bewölkungsgrade als klimatisches Element," *Archiv D. Seewarte*, XVI (1893), no. 5.

[2] A. Ångström, "Ueber die Schätzung der Bewölkung," *Met. Zeit.* (1919), pp. 257–262.

and frequency. A likely assumption is that cloudiness is proportional to the difference of the frequency of cloudy days minus that of clear days. If o means the average frequency of cloudy days and c that of clear ones, and n denotes the number of days of the month, season, etc., the following equations are valid for different regions:

Norway $$C = 51 + 51 \frac{o - c}{n}$$

Russian Empire $$C = 50 + 52 \frac{o - c}{n}$$

Palermo, Sicily $$C = 48 + 40 \frac{o - c}{n}$$

In Spain, the same definitions are used as in the United States. The formula reads:

Spain $$C = 50 + \frac{50}{n} (o - c)$$

TABLE 31   FREQUENCIES OF CLEAR (c) DAYS AND CLOUDY (o) DAYS AND THEIR RELATION TO CLOUDINESS (C%)

(*Example taken from the Observations at Blue Hill Meteorological Observatory*)

Month	$O - C$ Days	Cloudiness Obs %	Cloudiness Calc %	Obs. − Calc. %
Jan.	4	56	55	+1
Feb. .	2	52	51	+1
Mar..	5	56	57	−1
Apr. .	4	53	55	−2
May.....	.5	57	57	0
June .	5	56	57	−1
July	2	53	51	+2
Aug.	2	52	51	+1
Sept	1	49	49	0
Oct.	1	50	49	+1
Nov... .	4	54	55	−1
Dec . .	. 4	54	55	−1
Year	3	53	53	0

$$\Sigma(+) = 6$$
$$\Sigma(-) = 6$$

For Blue Hill Observatory, south of Boston, Mass. (42.2°N, 71.1°W, 640 ft) simultaneous observations of cloudiness and frequency (*c* and *o*) are available.[3]

With the assumption that $C = a + b (o - c)$, one obtains the equation $C = 47 + 2 (o - c)$. This equation differs from those previously mentioned in that no distinction is made between the different lengths of the months. This may be an example of avoiding an unfounded and exaggerated accuracy. The numbers in Table 31 show that the calculated values are in very good agreement with those observed, in spite of the simplified equation.

X. 2. FOG

The usual statistical characteristics are:

1) Average number of days with fog, divided sometimes into different degrees of density.
2) Probability of a foggy day (= the number of foggy days divided by the number of days in the month).
3) Average duration of fog, in hours, on a foggy day.

The last characteristic is computed by the method of random samples. The procedure is identical with that of the duration of precipitation which was discussed earlier (III. 10). Usually more detailed statistics are compiled only for the oceans. Six observations every 24 hours are made at sea, so that in this case the method of random samples is very successful.

[3] A. McAdie, "Observations and Investigations made at Blue Hill Meteorological Observatory, Mass , 1915 with Summaries for Thirty Years 1886–1915," *Annals of the Astronomical Observatory of Harvard*, vol. LXXIII, pt. III, (Cambridge, 1916).

CHAPTER XI

PRECIPITATION

IF THE PRECIPITATION is measured with the standard rain-gauge, not with a self-recording instrument, two primitive elements can be taken directly from the records:

1) Number of days with measurable precipitation [1]

    1a) with rain
    1b) with snow
    1c) with hail
    1d) with sleet

2) Total amounts of precipitation fallen during the single days, during a month, etc.

Starting from these data one proceeds to the following conceptions.

XI. 1. PROBABILITY (P) OF A DAY WITH PRECIPITATION

This is the quotient of the number of days with precipitation (r) over the total number of days of the period in question—month, season, year (n):

$$P = 100 \frac{r}{n} \%$$

(Boston, Mass., has 125 days with precipitation in the average of 60 years). Therefore

$$P = 100 \frac{125}{365.25} = 34.2\% = 342/1000$$

[1] Recently H Neuberger ("On the Measurement and Frequencies of Traces of Precipitation," BAMS, vol. 25, 1944, pp. 183–188) has published interesting statistics on traces of precipitation measured by means of new instruments which overcome some difficulties of the instruments of S P. Fergusson and others Here, it is not the instrumental side which is of interest but the fact that the investigations permit differentiating frequencies of traces and measurable precipitation Statistics of the two phenomena separated from one another would be valuable and also, e g., annual courses of the ratio of the frequencies of traces and real precipitation Not only applied climatology (agricultural) would profit much. Numbers of days with traces mean a new and important precipitation element if the traces are observed in a well-defined way. In any case, this element is at least as consequential as the number of days with dew.

If only per cent is calculated, it suffices to take the year with 365 days.

XI. 2. PROBABILITY THAT A DAY WITH PRECIPITATION IS A DAY WITH SNOWFALL

Generally, this probability should be calculated only for the average of the snowy period of a climatic region (Table 32), not

TABLE 32. PROBABILITY THAT A DAY WITH PRECIPITATION IS A DAY WITH SNOW-
FALL (P_s). BOSTON, MASS. NUMBER OF DAYS WITH PRECIPITATION AND SNOW-
FALL, (60 AND 31 YEARS RESPECTIVELY), TAKEN FROM G H. NOYES,
Meteorological Summary, Comparative Data
(U S. Weather Bureau, Boston, 1943)

Month	Number of Days with Precipitation	Number of Days with Snowfall	$\frac{Ps}{\%}$
July	10	0	0
Aug.	10	0	0
Sept.	9	0	0
Oct.	9	(<1)	(0)
Nov.	10	3	30
Dec.	11	8	73
Jan.	12	11	92
Feb.	10	10	100
Mar	12	7	58
Apr.	11	3	27
May	11	(<1)	(0)
June	10	0	0

for the whole year. If the number of days with precipitation is r, and s of them are days with snowfall, the probability

$$P_s = 100 \, s/r\%$$

XI. 3. AMOUNTS OF PRECIPITATION WITHIN CERTAIN PERIODS

The amounts of precipitation for single days are totaled for different sections of the year according to the purposes in question. There are average sums and individual amounts per week, ten days, month, season, year. For agricultural purposes, the amount of precipitation which occurs during the vegetative period (growing season) will be of interest (see the explanation in the chapter on temperature, VI. 3). The beginning and end of these periods is either derived from the annual course of temperature, or from biological and phenomenological experiences; for instance, the time from the average date of sowing to that of the beginning of the flowering season.

In an ordinary climatography, at least the monthly and the annual amounts have to be given. Average amounts of precipitation should be given in millimeters; frequently better, in centimeters; or else, if possible, in hundredths of inches and otherwise, in tenths. This is the accuracy which corresponds to the exactness of the observations.

XI. 4. THE ANNUAL COURSE OF THE AMOUNTS OF PRECIPITATION

The annual amount of precipitation *may* be considered as being evenly distributed over all days of the year. Monthly sums can then be calculated. The resulting course shows, however, variations caused by the unequal length of the months. This inconvenience gives rise to different methods for eliminating this source of error.

XI. 4 a. *The Reduction to Months of Equal Length*

This can be achieved in two ways. The simpler way is to reduce the monthly amount to equal months of 30 days. The amounts for the months with 31 days are reduced by 3.2%. The amount for February (28.25 days) is increased by 6.2%. Table 33

TABLE 33. MONTHLY PRECIPITATION (MM) AT A PLACE IN CENTRAL EUROPE REPRESENTED BY DIFFERENT METHODS

1	2	3	4	5	6	7
Month	Average monthly precipitation mm	Col 2 reduced to equal months of 30 days mm	Col 2 in thousandths of the annual amount	Even distribution of an annual amount of 1000	Écart pluviométrique relatif (Col 4 minus Col 5)	Relative pluviometric coefficient
Jan. .	37	36	57	85	−28	0 67
Feb. .	33	35	51	77	−26	0.66
Mar. .	47	45	73	85	−12	0 86
Apr .	53	53	82	82	0	1 00
May	71	69	110	85	25	1.29
June. .	70	70	108	82	26	1.32
July..	79	76	122	85	37	1 43
Aug..	69	67	106	85	21	1 25
Sept	50	50	77	82	− 5	0 94
Oct. . ..	47	45	73	85	−12	0 86
Nov	45	45	69	82	−13	0 84
Dec.	47	45	73	85	−12	0.86
Year.....	.648	(636)	1000	1000		

contains in the second column the original monthly sums at a place in Central Europe. They are reduced to months of equal length of 30 days in the third column.

The second method is to reduce the original sums (column 2) to a month which has the length of the twelfth of a year. That is 30.438 days. Therefore the amounts

for February have to be multiplied by 1.077
for the months with 30 days,   by 1.015
for the months with 31 days,   by 0.982.

XI.   4 b. *Amount of Precipitation on an Average Day of the Month*

The monthly amount is divided by the number of days of the month in question. The advantage of this method is that the inequality of the months is eliminated. The great disadvantage however, is, that these numbers are absolutely fictitious. Thus, it was not of much value for F. H. Bigelow [2] to calculate "Daily Normals of Precipitation" for a great number of stations in the United States. Precipitation is always a discontinuous climatological element. Every day and everywhere there is always a temperature; but there are days with precipitation and days without. An evenly interpolated distribution gives a wrong picture of the phenomenon.

XI.   4 c. Écart Pluviométrique Relatif *and the* Relative Pluviometric Coefficient

A. Angot suggested two other methods of representation, which though not very illuminating have been frequently used in climatological papers and books. His *Écart pluviométrique relatif* may best be translated *relative monthly deviation*. In Table 33, column 5, the numbers of an even distribution over the months are shown for an annual amount of 1000. In column 4, the actual distribution per mille is calculated from column 2. In column 6, the differences (column 4 minus column 5) are given. These differences are called *écart pluviométrique relatif*, or relative monthly deviations.

Monthly quotients are calculated with the numbers of column 4 over the homologous numbers in column 5. These ratios are called

[2] Frank H Bigelow, "The Daily Normal Temperature and the Daily Normal Precipitation of the United States," U S Weather Bureau, *Bulletin R* (Washington, D. C., 1904).

relative pluviometric coefficients (column 7).[3]   Both expressions are scientifically well founded and eliminate the inequalities of the months.

All in all, the reduction to equal months of 30 days seems to be the best expedient.   The small difference between the actual annual sum and the reduced sum (see last line of columns 2 and 3) is of no significance.

XI. 5. ABSOLUTE AND RELATIVE VARIABILITY. QUOTIENT OF VARIATION. METHOD OF RANDOM SAMPLES APPLIED TO RECORDS OF RAINFALL [4]

These three topics have been discussed in connection with related problems of mathematical statistics. (See III. 9 and III. 10).

XI. 6. RAIN INTENSITY (I)

The ratio between the total amount during a given period (A) and the number (N) of days with precipitation is called *rain intensity*.   It is the average amount of rain of one rainy day.

$$I = \frac{A}{N}$$

In a 50–60 years' average for August, the amount of rain at Boston, Mass., is $A = 3.62$ inches, and the number of rainfall days is $N = 10$.   Therefore the rain intensity of August at Boston is $I = 3.62/10 = 0.36$ inches $= 9$ mm per rainy day.   The corresponding data for Cherrapunji in the Khasi Hills in British India (25.3°N. 91.8°E), in June, the month of maximum, are: $A = 2632$ mm; $N = 24.8$; $I = 106$mm per rainy day.   Cherrapunji is located in one of the rainiest regions of the world.   There rain intensity in the wettest month is twelve times greater than at Boston.

Rain intensity is a very significant characteristic of a climate.

[3] Dr. C. F. Brooks kindly draws my attention to two papers of B C. Wallis, "The Rainfall of the Northeastern United States" and "The Distribution of the Rainfall in the Eastern United States," MWR, vol. 43 (1915), pp. 11–14 and 14–24, where the *pluviometric coefficient* is applied to the rainfall in the eastern United States.

[4] Following a British custom, *rainfall*, being the shorter expression, is often used instead of *precipitation*.

XI. 7. WET AND DRY SPELLS

The average and extreme lengths of wet and dry spells are of decisive influence upon organisms. A wet spell is defined as a period of consecutive days with at least 0.01, 0.05, etc., inches, or 0.1, 0.2, 0.5, etc., millimeters of rain. The thresholds on which the definition is founded have to be expressly indicated because no general agreement exists in this respect.

It is assumed that about ten years' records are at hand. This is sufficient. A much shorter period should not be chosen. If the ten years are available, the length of the spells (wet or dry periods) should be excerpted from the monthly records. An isolated rainy day means a spell of one day for statistical reasons. Then

    a) the average length of a wet (dry) spell
    b) the average length of the longest wet (dry) spells
    c) the extreme lengths

are computed for each month. If a spell overlaps from one month to the next, or if its length is greater than one month, the rules for hot and cold spells govern (see VI. 4).

A dry spell means a period of at least five consecutive rainless days. It goes without saying that the establishment of a minimum limit is necessary; a shorter period is unlikely to damage the vegetation in any way. "Dry periods" longer than five days are not considered as being interrupted by a rain less than 1mm (<0 04 inch).

On the coast of the middle Adriatic Sea, dry spells up to the extreme of 81 days are recorded in summer. The longest rainy period observed in the eastern Alps within 40 years had a duration of 16 days (in September). These two examples serve to illustrate dry and wet spells.

The definitions mentioned are by Julius Hann, slightly modified by the author. It is regrettable that there is no international definition of wet and dry spells.

In the British Meteorological Service, other definitions are used:[5]

Drought is dryness due to lack of rain. Certain definitions have been adopted in order to obtain comparable statistical information on the subject of droughts. An *absolute drought* is a period of at least 15 con-

---

[5] *The Meteorological Glossary* (3rd ed., Meteor Off London, 1939). See pp. 68, 161, 212.

secutive days, to none of which is credited 0 01 in. of rain or more. A *partial drought* is a period of at least 29 consecutive days, the *mean daily* rainfall of which does not exceed 0.01 in.; (the word mean is to be emphasized). A *dry spell* is a period of at least 15 consecutive days to none of which is credited 0.04 in. rain or more.

Rain spell is a period of at least 15 consecutive days to each of which is credited 0.01 in rain or more; thus the definition of the term rain spell is analogous to that of the term "absolute drought.
Wet spell. Analogous to the term dry spell, the wet spell is a period of at least 15 consecutive days to each of which is credited 0.04 in. of rain or more."

These definitions are surely well fitted to the climate of the British Isles. The definitions by Hann-Conrad have perhaps more general applicability and are less complicated.

The American definition of drought reads: "period of 20 (or 30) consecutive days or more without 0.25 inch of precipitation in 24 hours during the season, March to September, inclusive."

XI. 8. CHARACTERIZATION OF THE HYDROMETEORIC TYPE OF A PERIOD

Frequently it is necessary to give a general characterization of a week, month, year, etc., as far as rainfall is concerned. The question as to whether a month is to be called dry, for instance, or extremely dry, wet or extremely wet, etc., in a given climate, is answered in III. 3. By means of the standard deviation σ, the problem is solved in a well-defined and reproducible way.

XI. 9. SNOW

Table 34 shows a form representing snow conditions which has proved to be advantageous. The form serves for a single year as well as for a series of years. In the latter case, averages replace the individual values. If possible, extreme values taken from the period in question should be inserted; e.g., between columns 6 and 7, the earliest and the latest date of both the beginning of the first and the ending of the last snow-cover. This makes 4 columns more, with increased inconvenience for printing. A supplementary table of extremes is often more convenient.

In Column 7 is shown the number of days on which a snow-cover actually exists.

In contradistinction to this characteristic, the *season of snow-cover* has to be indicated in Column 8. The season means the number of days between the date of the beginning of the first snow-cover and that of the end of the last one. Within the "sea-

TABLE 34   FORM FOR TABULATING SNOW CONDITIONS FOR A NUMBER OF STATIONS

Period   .... to ... . (.   years)

[Remarks regarding observation-hour and methods]

1	2	3	4	5	6	7	8	9	10	11	12
				Date of the							
Name of place (Country)	Altitude (m, feet)	Geographical latitude	Geographical longitude	Beginning of the first	End of the last	Number of days with snow cover	Season of snow cover (days)	Date of the first snowfall	Number of days with snowfall	Sum of the depths of freshly fallen snow (cm, inches)	Maximum depth of snow on ground (cm, inc.)
				Snow cover							

son" the snow-cover can melt away and be formed anew. The number of days with an actual snow-cover is called *duration of snow-cover*, and is always smaller than, or at the most, equal to the season. The greater the elevation and the geographical latitude, the smaller the difference between season and duration. For example, in the eastern Alps at one place at 4000 feet altitude the difference is 20 days, at another place at 700 feet it is 52 days.

The data of column 11 can be given only if the freshly fallen snow is regularly measured. These numbers are totaled for the month or the year.

The records of the coöperative stations of the U. S. Weather Bureau enable one to compile statistics of the lengths of the single periods of the snow-cover. This is important especially for regions where the snow surface offers easy transportation, e.g., logging, skiing, etc.

The variability of the duration as well as of the season is of importance and of interest. It decreases with elevation under otherwise equal conditions. Average values of duration and season have to be reduced to a standard period by the method of differences. Values derived from different periods are not comparable.

If Z means the duration in days and S the annual amount of melted snow in millimeters or inches, then the ratio $Q = Z/S$ yields the duration value (Q) in days for 1mm (inch) of snow. This expression can be useful in some investigations.

The percentage of the total amount of precipitation which falls in the form of snow should be indicated in every regional description of snow conditions.

In more detailed studies into the depth of snow on the ground, average month or half-month values are calculated. The statistics of longer series of years yield probabilities of certain average depths of snow in certain weeks, half months, etc.[6]

In mountainous countries, the depth of snow on the ground is, of course, dependent to a high degree upon altitude. Therefore, also, cartographic representations of depth, duration, season, etc., can be based only on the principle of isanomals. Further details are beyond the scope of the present work.[7]

[6] For a good pattern see R. G Stone, "The Distribution of the Average Depth of Snow on Ground in New York and New England: Method of study," *Transactions of the Am. Geophysical Union, 19th Annual Meeting* (1938), pp 486–492, and the second part, "Curves of Average Depths and Variability," *ibid.* (1940), pp. 672–692.

[7] For further discussion see "Mitteilungen uber die Schneedecken-Verhältnisse in den Ostalpen," in Gerland's *Beitrage zur Geophysik*, as follows: (1) V. Conrad and M. Winkler, "Beitrage z. Kenntnis d. Schneedecken-Verhaltnisse," vol. 34 (1931), p. 473; (2) V. Conrad, "Die Schneedeckenzeit, ihr Anfangs-und Enddatum in den Ostalpen," vol. 43 (1934) p. 225; (3) V. Conrad, "Der Anteil des Schnees am Gesamtniederschlag und seine Beziehungen zu den Eiszeiten," vol. 45 (1935), p. 225; (4) F. Steinhauser, "Ueber den Schneeanteil am Gesamtniederschlag im Hochgebirge der Ostalpen," vol 46 (1936), p 405; (5) E. Ekhart, "Die Andauer der Schneedecke nach Stufenwerten der Schneehöhe," vol. 50 (1937), p. 184; (6) V. Conrad and O. Kubitschek, "Die Veränderlichkeit und Machtigkeit der Schneedecke in verschiedenen Seehohen," vol. 51 (1937), p. 100.

Additional Remark: According to an investigation in a rather extensive material, the variability of the depth of snow on the ground is so great that a reduction to a given period is not successful Therefore, the discussion only of synchronous series gives reliable results.

XI. 10. INDEX OF ARIDITY

This concept represents an element combined of hydrometeoric elements and temperature elements. First, E. de Martonne's index may be mentioned (in a modified form):

$$I = \frac{nr}{t + 10}$$

where

- n = the average number of rainy days within the period in question,
- r = the average rainfall in the same period,
- t = the average temperature.

The three variables are related to a certain month, a season, or a year.

Another index of aridity has been established by W. Gorczynski.[8] His aridity coefficient (AC) considers the influence of latitude, of the range of temperature, and E. Gherzi's measure of variation of precipitation $(M - m)/\bar{p}$. (See III. 9.) The fundamentals of the AC are, however, not yet so well investigated that its routine use can be recommended.

Generally, such indices may be applied only with great caution. Often the results are satisfactory in one region and show contradictions in another. Furthermore, the boundary conditions are sometimes not fully considered, so that further investigations appear desirable.

The numerical indices calculated from these formulae yield relative values fitted only for *comparison* of the conditions at different stations. Other similar attempts (*viz.*, continentality, border lines of climatic types) are discussed in part XVII. 1 and 2.

XI. 11. RAIN-HISTOGRAM

Two methods of graphical representation of the annual course of rainfall are usual:

1) the histogram
2) the smooth curve.

Both kinds are known from the representation of the frequency distribution (see Fig. 4). The histogram originates from the idea

[8] *The Aridity Coefficient and its Application to California* (The Scripps Institution of Oceanography, La Jolla, Calif., 1939).

already mentioned that precipitation is a discontinuous element. Consequently, this representation is logical. The disadvantage of the method is the difficulty of drawing, and the inaccuracy resulting therefrom. On the other hand, nobody can forget or overlook the fact that rainfall is a discontinuous element; altogether the smooth curve is preferable. It is more easily drawn and better comparable with other elements.

PART III

METHODS OF SPATIAL COMPARISON

CHAPTER XII

COMPARISON OF OBSERVATIONAL SERIES OF DIFFERENT PLACES. GEOGRAPHICAL DISTRIBUTION OF CLIMATOLOGICAL ELEMENTS

XII. 1. Uniform Tendency of the Variations of Average Weather over Large Regions. Consequences Drawn from This Tendency for Comparing Series of Elements Observed Simultaneously at Different Places

IN THE BEGINNING of the first part of this work, some remarks were made about different kinds of observations of climatological elements, as far as they are of methodological interest. It was there demonstrated (I 3. a.) that a change of instrument, even if it means an important improvement, can cause a break in a climatological series.

Another very trivial example may be added here. At a "secular" station, the rain gauge happened to leak after having been used for some years. The hole in the bottom was very tiny in the beginning, so small that it closed by itself when dust and dirt were caught with the rain. Then for a time the gauge was watertight. It began to leak again when the dirt was washed away by heavier rain. The acids contained in the rainwater widened the hole in the course of time till the vessel leaked so much that the observer became aware of the defect. Then the gauge was repaired or exchanged, and the cycle began anew. The lifetime of the gauge was determined by workmanship and material.

This is the story of one of the many thousands of rain gauges in the world. And what does it mean for climatological methods?

As long as the rain gauge is intact, the annual amounts may fluctuate around a certain mean. This decreases when the gauge leaks, increases a little when the hole is nearly closed by chance, decreases again, and so on. The mean happens to reach its former value in the period after a new gauge is installed. Thus, a false cycle of a semi-periodical variation of rainfall comes into existence. This simple and trivial story is obviously of great importance to the student of periodicities in long climatological series. No doubt in this regard many a scientist has been misled.

A series of observations such as the one described, showing variations caused by unnatural influences, is called *inhomogeneous.* It is too easy to find manifold causes of similar inhomogeneities to make further explanation necessary here (growing trees, new buildings close to the shelter, another observer, etc.). Accordingly, the following definition is offered: *A numerical series representing the variations of a climatological element is called "homogeneous" if the variations are caused by and only by variations of weather and climate.*

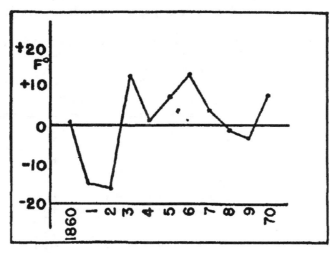

FIG. 26.   Variations of January temperature at Moscow in the years 1860–1870.
(Deviations from the normal, F°)

Obviously, it must be the first step of the climatologist to examine whether or not the series is homogeneous, in order to avoid inferences which are strictly false. This examination encounters a great difficulty. There are real climatic variations from year to year. The example of Figure 26 shows the variations of average January temperatures at Moscow in the years 1860 to 1870. We know that the observations are incontestable. If we did not know this fact, we should not be able to indicate whether the increase of temperature, e.g., from 1862 to 1863, was caused by nature or otherwise. Therefore we have to conclude that it is not possible to discriminate between natural and artificial variations of an element if the observations of only one place are available.

In other words, it is generally impossible to decide whether or not a series of observations is *absolutely homogeneous*. On the other hand, it is clear that the variations of weather and climate are not restricted to a single place, with the exception of a cloudburst over one spot, or kindred phenomena. This principle becomes better exemplified, of course, if averages over a month, a season, etc., are considered. In longer periods, local extreme deviations are compensated and the trend of the variation is maintained over large regions.[1]

First, temperature may be taken as an example. New York City and New Haven, Connecticut, are 80 miles apart. The January temperatures of the ten years reproduced in Table 35

TABLE 35   The Quasi-Constancy of the Differences between January Temperatures at Two Places, 80 Miles Apart

New York City (40 7°N, 74.0°W, 314 ft) (Y)
New Haven, Conn. (41 3°N, 72 9°W, 106 ft) (H)

ΔY = Deviations Y ΔH = Deviations H

(1) Year	(2) Y, °F	(3) H, °F	(4) $(Y - H)$°F	(5) ΔY, F°	(6) ΔH, F°	(7) $\Delta(Y - H)$F°
1911	34 5	32.1	+2 4	+3.5	+2.9	+0 7
1912	23.2	21.2	2 0	−7.8	−8.0	+0.3
1913	39.7	38 2	1.5	+8.7	+9 0	−0 2
1914	31 1	29 3	1.8	+0.1	+0 1	+0.1
1915	33 8	33 0	0.8	+2.8	+3.8	−0 9
1916	35 1	33 0	2 1	+4 1	+3.8	+0.4
1917	32.1	30 6	1 5	+1 1	+1 4	−0.2
1918	21 3	19 8	1 5	−9 7	−9.4	−0.2
1919	34.9	33 8	1.1	+3.9	+4.6	−0 6
1920..	23 8	21.4	2 4	−7 2	−7.8	+0.7
Mean	31 0	29.2	1 7			
$\Sigma(+)$				24 2	25 6	2.2
$\Sigma(-)$				24 7	25.2	2 1
$v = \pm..$				4 89	±5 08	±0.43

show great variations. The temperature in New York varies—by chance—by the same amount as in New Haven, i.e., 18.4 F°. The differences vary by 1.6 F°.

[1] On the other hand, it is very unlikely that the same kind of artificial influences occur synchronously at two different places.

The mean deviations (average variabilities) are (column 5, 6):

$$\overline{\Delta Y} \doteq \pm 4.9; \qquad \overline{\Delta H} \doteq \pm 5.1$$

where Y means New York and H means New Haven.

The average variability of the differences is, however:

$$\Delta(Y - H) = \pm 0.4$$

Thus, the differences of the temperatures can be called *quasi-constant*, because their variations are the twelfth part of those of the actual temperatures. Similar conditions are clearly demonstrated in the example shown in Table 36A,[2] for precipitation, even

TABLE 36A. THE QUASI-CONSTANCY OF THE RATIOS BETWEEN RAINFALL IN JULY AT TWO PLACES, 11 MILES APART

Boston, U. S. Weather Bureau (42 3°N, 71 1°E, 124 ft) (B)
Waltham, Cambridge Water Works (20 ft) (W)

(1) Year	(2) Boston (B) in	(3) Waltham (W) in	(4) B/W	(5)* ΔB	(6)* ΔW	(7)* $\Delta(B/W)$
1911	4.65	4.83	0 96	+0.69	+0 73	−0 02
1912	5 16	4 38	1 18	+1.20	+0 28	+0 20
1913	2 69	3 29	0 82	−1 27	−0 81	−0 16
1914.	2 64	2 46	1 07	−1 32	−1 64	+0 09
1915	8 85	9 87	1 07	+4 89	+5 77	−0 08
1916	5 67	5 31	0 95	+1 71	+1 21	+0 09
1917	1 10	0 93	1 18	−2 86	−3 17	+0 20
1918 ..	2 64	2.99	0 88	−1 32	−1.11	−0.10
1919 .	4 63	5.21	0 89	+0 67	+1.11	−0 09
1920 ..	1.56	1.78	0 88	−2.40	−2.32	−0 10
$\bar{m}$ ..	3.96	4.10	0.98			
$\Sigma(+)$				9 16	9.10	+0 58
$\Sigma(-)$				9 17	9 05	−0 55
$v = \pm$				1 83	1 82	0 11

$$v_r(B,W) = \pm 45\%\dagger \qquad v_r(B/W) = \pm 11\%$$

* Deviations from the means of the period 1911 to 1920
† $v_r(B,W)$ denotes the average relative variability of the amounts at B and W.

if here for this element a much smaller distance between the two places has been chosen. If precipitation is concerned, it is not

[2] Temperature data are taken from H Helm Clayton, *World Weather Records* (Smithsonian Miscellaneous Collections, vols. 79, 90), precipitation data from X. H. Goodnough, "Rainfall in New England," (*Journal of the New England Water Works Association*, Boston, June 1930.)

the differences that count, but the ratios between the amounts at the two places. The explanation will be given later on.

Here, the ratios are much less variable than the rainfall itself, as is indicated by the mean deviations (v) at the bottom of columns 5, 6, 7. In reality, this behavior of the variability of the ratios is not yet decisive. In this example, it is a question of the amounts, which means that relative, not absolute, variabilities (v_r) have to be considered. (See III. 9.) Therefore, the absolute variabilities must be divided by the respective arithmetic means (m). These relative variabilities (v_r) read:

$$v_r(B, W) = 45\%$$

and

$$v_r(B/W) = 11\%$$

So it is clear that in the example the ratios between the amounts can be assumed as quasi-constant, in comparison with the actual rainfall amounts.

The quasi-constancy of the ratio B/W can also be shown by another very common representation. In Table 36B, the amounts

TABLE 36B   FLUCTUATIONS OF RAINFALL AMOUNTS AS PERCENTAGE OF THE AVERAGE VALUES AS ILLUSTRATIVE (AFTER L W POLLAK) OF THE QUASI-CONSTANCY OF B/W

Year	Percentage of average 1911–1920 Boston	Waltham	Ratio B/W	$\Delta\dfrac{B}{W}$ units of second decimal place
	%	%		
1911	117	118	0 99	− 3
1912	130	107	1 21	19
1913	68	80	0 85*	−17
1914	67	60	1.12	10
1915	*223*	*241*	0 93	− 9
1916	143	130	1 10	8
1917	28*	23*	*1 22*	20
1918	67	73	0.92	−10
1919	117	127	0.92	−10
1920	39	43	0 91	−11
			1 02	
m	3.96 in.	4.10 in	0 97	11.7

of rain in B and W are given as percentage of average, 1911 to 1920. The amounts of rainfall at Boston vary from 28% to 223% or by 195%, and those for Waltham by 218%, while the ratios of B/W vary by 37% only.

These examples show that the variations of *average* weather particularly have identical tendencies over rather large regions. If the temperature drops at a place *a* it may drop too at the place *b* (not too far distant from *a*), and the same is true for rainfall, cloudiness, etc.   In other words, there is no reason why variations of differences or ratios should be systematically influenced by the average weather events.

XII. 2. CRITERIA OF RELATIVE HOMOGENEITY

As previously stated, a series of numbers representing the variations of a climatic element is called *homogeneous* if the variations are caused by and only by climatic influences.   Now we have shown that these cannot give rise to *systematic* variations of "differences" or "quotients."

The two facts together yield an exact definition of *relative homogeneity*, as follows: *A climatological series is relatively homogeneous with respect to a synchronous series at another place if the differences (ratios) of the pairs of the homologous averages represent a series of random numbers which satisfies the law of errors.*

To determine whether or not two series are relatively homogeneous is now a problem of statistical-mathematics.   I want to give two criteria,[3] those of Helmert and Abbe.

XII. 2. a. *Helmert's Criterion*

The criterion of F. R. Helmert, well known in geodesy, is applicable to the present problem and offers a first estimation.   If two consecutive deviations of the series have the same sign, we speak of a *sequence*.   The number of sequences $(+ \ + \ \text{or} \ - \ -)$ is called S.   A change of sign of two consecutive elements is called *change* $(+ \ - \ \text{or} \ - \ +)$.   The number of changes is called C. Now it may be assumed that the elements of a series are not systematically influenced but represent random numbers.   Deviations are calculated.   The consecutive signs yield S sequences and C changes.   Then Helmert's criterion indicates that

$$S - C = 0$$

with a standard deviation

$$\pm \sqrt{n - 1}$$

[3] V. Conrad, "Homogenitätsbestimmung meteorologischer Beobachtungsreihen," *Met Zeit.*, 1925, p. 482.

where n equals the number of elements. In other words: in the absence of a systematic influence

$$- \sqrt{n-1} \leqq (S - C) \leqq + \sqrt{n-1}$$

In the example of Table 35, the number

$$n = 10$$

therefore $\sqrt{n-1} = 3$ and

$$- 3 \leqq (S - C) \leqq + 3$$

Counting sequences and changes in column 7, we get:

$$S - C = 3 - 6 = -3$$

Thus, Helmert's criterion shows that the series of differences may not be systematically influenced. This means that the two series of January temperatures for the years 1911 to 1920, one taken at New York, the other at New Haven, are relatively homogeneous.[4]

XII   2. b. *Abbe's Criterion*

Another criterion, that of Carl Abbe, which originated from physical research, has been suggested by the author.[5]

There may be a series of deviations from an arithmetical mean:

$$d_1, d_2, d_3, \cdots d_i \cdots d_n$$

From this series, two other series are derived:

1) $A = d_1^2 + d_2^2 + \cdots d_i^2 + \cdots d_n^2 = \sum_{i=1}^{i=n} (d_i)^2$

2) $B = (d_1 - d_2)^2 + (d_2 - d_3)^2$
    $+ \cdots (d_i - d_{i+1})^2 \cdots (d_{n-1} - d_n)^2 + (d_n - d_1)^2$

$$= \sum_{i=1}^{i=n} (d_i - d_{i+1})^2$$

Series B is considered as cyclic, so that the last term is formed as the difference of the last and the first deviation. By this proce-

---

[4] Here a series of only ten elements is given, for the sake of brevity. Generally, the numbers of elements should be rather greater to permit application of probability criteria. See F R. Helmert, *Die Ausgleichsrechnung* . . . (2nd ed., Leipzig, 1907), p. 333.

[5] V. Conrad, *Met. Zeit*, 1925.

dure, the numbers of items of the two series A and B are equal. Series B reads:

$$B = \sum_{i=1}^{i=n} (d_i - d_{i+1})^2 = \sum_{i=1}^{i=n} (d_i)^2 + \sum_{i=1}^{i=n} (d_{i+1})^2 - 2 \sum_{i=1}^{i=n} d_i d_{i+1}$$

Now

$$\sum_{i=1}^{i=n} (d_i)^2 = \sum_{i=1}^{i=n} (d_{i+1})^2$$

as identical values. Therefore

$$B = 2 \sum_{i=1}^{i=n} (d_i)^2 - 2 \sum_{i=1}^{i=n} d_i d_{i+1}$$

or

$$B = 2A - 2 \sum_{i=1}^{i=n} d_i d_{i+1} = 2A - F$$

The most probable value of

$$F = e_1 e_2 + e_2 e_3 + \cdots + e_{n-1} e_n + e_n e_1 = 0$$

if the deviations are random numbers and n is a great number. Under these conditions one can imagine that it is always possible to combine any product $\pm e_i e_{i+1}$ with another term of equal absolute magnitude and opposite sign. Consequently, the series F determines whether or not there is a systematic influence. If there is none, or in other words, if the deviations are random numbers, $F = 0$ in so far as n is a large number.

In concrete cases, F only approximates the value of zero with increasing n so that

$$\frac{2A}{B} = 1 \pm \sqrt{\frac{1}{n}}$$

(Appendix II contains the values of $1/\sqrt{n}$ for $n = 1$ to 119, so that the limits can be easily determined.) This is Abbe's criterion, which is fulfilled if the series in question is not systematically influenced.

When we have the problem which arises from the series of differences or quotients, respectively, the application of Abbe's criterion is most valuable because it considers the sequence of the signs as well as the quantity of the single deviations.[6] Tables 37

[6] A third criterion regarding groups of equal signs is given by V Conrad, "Die klimatologischen Elemente," in Köppen-Geiger, *Handbuch der Klimatologie*, vol. I B, p. 114.

TABLE 37.   ABBE'S CRITERION APPLIED TO A SERIES OF JANUARY TEMPERATURES AT
NEW YORK (Y) AND NEW HAVEN (H)
(See Table 35, Column 7)

Year	$d_i = \Delta(Y - H)F°$	$\lvert d_i - d_{i+1} \rvert$	d_i^2	$(d_i - d_{i+1})^2$
1911 .	+0 7	0 4	49	16
1912. . .	. +0.3	0.5	.09	.25
1913 .	−0 2	0 3	.04	.09
1914	+0 1	1 0	01	1 00
1915	−0 9	1.3	81	1.69
1916	+0 4	0 6	16	36
1917..	−0 2	0.0	04	00
1918 .	−0 2	0 4	.04	.16
1919	−0 6	1 3	.36	1 69
1920.	+0 7	0 0	.49	.00
			$A = 2\ 53$	$B = 5\ 40$

TABLE 38   ABBE'S CRITERION APPLIED TO THE RAINFALL SERIES OF TABLE 36A*

(1) Year	(2) $d_i = \Delta(B/W)$	(3) $\lvert d_i - d_{i+1} \rvert$	(4) d_i^2	(5) $(d_i - d_{i+1})^2$
1911	−0 01	22	0001	.0484
1912 .	+ 21	39	0441	.1521
1913	− 18	28	0324	.0784
1914 .	+ 10	17	0100	0289
1915	− 07	05	.0049	.0025
1916	− 02	23	0004	0529
1917	+ 21	30	0441	.0900
1918 .	− 09	01	0081	.0001
1919 .	− 08	01	0064	0001
1920.	− 09	08	0081	0064
			$A = 1586$	$B = .4598$

* The series of deviations in column 2 of Table 38 differs slightly from column 7 of
Table 36A.   The result is valid for both series because that of Table 36A gives

$$2A/B = 0.77$$

and 38 give examples of the application of Abbe's criterion.   Here
the series of differences (column 7 of Table 35) and the series of
quotients (column 7 of Table 36A) are examined for relative
homogeneity.

From Table 37 may be calculated:

$$\frac{2A}{B} = \frac{2 \times 2.53}{5.40} = 0.94$$

As $n = 10$, $1/\sqrt{n} = \pm 0.32$ (See Appendix II).   Therefore

$$0.68 \leqq \frac{2A}{B} \leqq 1.32$$

if no systematic influence exists.   The value $2A/B$ lies indeed between these limits.   Therefore it is probable that the two series are relatively homogeneous.

The criteria of Abbe and Helmert arrive at the same result. From Table 38 we derive:

$$\frac{2A}{B} = \frac{2 \times 0.1586}{0.4598} = 0.69$$

and because

$$0.68 \leqq \frac{2A}{B} \leqq 1.32$$

Abbe's criterion is fulfilled and there is no reason to assume a systematic influence upon the $\Delta(B/W)$—series.[7]

Counting the sequences (S) and the changes (C) of column 2 (Table 38) it follows that: $S = 3$, $C = 6$, and $S - C = -3$.

As $\sqrt{n - 1} = 3$, Helmert's criterion is fulfilled if

$$-3 \leqq (S - C) \leqq +3$$

and again the two criteria agree with one another.

If the relative homogeneity between the series of an element at two places a and b is given, and a third station c, suitably located, is available, a conclusion can be drawn regarding the absolute homogeneity of one of the series.   If the pairs of series (a, b), (b, c), and (c, a) are relatively homogeneous, it is rather probable that each of the three series is absolutely homogeneous.   Thus a sort of climatological triangulation can be accomplished.   The larger the number of places included, the greater is the likelihood that the conclusions regarding absolute homogeneity will be true.

The chief trouble with all these investigations is the scarcity of stations.   The meshes of the climatological networks are generally too wide, so that the application of the higher and exact methods meets with difficulties.   Nevertheless, the examination of a pair of records for relative homogeneity is fundamentally important if the climate of a region is to be described accurately. Only by this examination can the natural climatic changes be separated from the unnatural by means of an exact analysis.

One should never forget that a leaking gauge can counterfeit the finest periodicity.

[7] It may be emphasized that Abbe's criterion, in the form discussed here, *cannot be applied to smoothed series*   See V. Conrad and O. Schreier, "Die Anwendung des Abbe'schen Kriteriums auf geophysikalische Beobachtungs Reihen," Gerland's *Beiträge zur Geophysik*, XVII (1927), 372.

XII. 3. REDUCTION OF CLIMATOLOGICAL AVERAGES TO A CERTAIN PERIOD

XII. 3. a. *Method of Differences*

Hitherto it has been assumed that the observations at the various places are operated synchronously. Then the comparableness is guaranteed as long as the series are mutually relatively homogeneous.

TABLE 39 TEMPERATURE CONDITIONS OF DIFFERENT PERIODS AT ST. PAUL, MINNESOTA (45 0°N, 93.0°W, 837 FT) IN JANUARY

(1) Year	(2) January °F	(3) Year	(4) January °F	(5) Col 4 — Col 2
1883	0 9	1898	22 8	+21 9
1884	7 5	1899	13 8	6 3
1885	4 3	1900	21 2	16.9
1886	3 7	1901	16 3	12 6
1887	0 7	1902	18 7	18 0
1888	−1.1	1903	15.3	16 4
Mean ..	. 2 7		18 0	15 3

Series of different periods, however, are not comparable with one another. A short example is given in Table 39.[8] If one did not know that the two series (columns 2 and 4) originated in the same place, one might suppose that the series in column 4 represented the temperatures at a place 8° latitude farther south. In addition, the July temperatures are higher in the period of cold winters, and not only are the actual temperatures very different in the different periods, but also the annual ranges.

Another example (Table 40) gives the amounts of rainfall during different periods at Malden Island in the South Pacific.

TABLE 40 RAINFALL AT MALDEN ISLAND (4°S, 155°W) IN DIFFERENT PERIODS

Period	Average amount inches
Years 1890 to 1899	16.30
Years 1911 to 1920 .	41.97
Januarys 1890 to 1899	0 55
Januarys 1911 to 1920 .	. 5.55

[8] Data drawn from R. DeC. Ward and C F. Brooks, "The Climates of North America," Köppen-Geiger, *Handbuch der Klimatologie*, pt. J. (1936).

There, to the best of the author's knowledge, the greatest known contrasts occur.

From these examples, it is clear that the changes in the climatic elements with time are so effective that a comparison between average values of temperature, rain, cloudiness, etc., at different places can lead to a reasonable conclusion only if identical periods are considered. In view of the scarcity of climatological data, however, the restriction of comparisons to a certain period would seriously impede climatographic work. The solution of this fundamental problem is found in the quasi-*constancy* of differences (quotients) of synchronous averages of different climatic elements, which has been discussed earlier. We take two more examples from Table 35:

January	New York (Y)	New Haven (H)	$(Y + H)/2$	$(Y - H)$
1916	35.1°F	33 0°F	34 1	2 1°F
1918	21 3	19 8	20.5	1 5
Diff. .	13.8	13.2	13.6	0 6

This means that the January temperature changed by 13.6 F° from 1916 to 1918, but the *difference* between Y and H changed by only 0.6 F°. This behavior has been called "quasi-constancy" of the differences.

Another example from Table 36A, shows the following conditions:

RAINFALL, IN INCHES

	Boston	Waltham	Ratio B/W
1915 July	8 85	9 87	0 90
1917 July	1.10	0.93	1.18

In other words, while the rainfall was 9 times greater in the wettest July (1915) than in the driest July (1917) the ratio of the rainfall in Boston over that in Waltham changed by only 24%.

Thus the quasi-constancy of differences (quotients) is evident, both by means of a general geophysical conception and by practical examples.

A climatological station S [9] has been operated N years, where N is a relatively large number, perhaps greater than 25.

A second station a has been operated n years. Let

$$n < N \quad \text{and}$$

[9] The letter S (Secular Station) has, of course, another meaning from that in the previous paragraph dealing with Helmert's and Abbe's criteria.

the n years lie within the period of N years. The station S is called a "normal" or "secular" station. Let the n-year average temperature at the station a be

$$i(a, n)$$

and the same n-year average at the station S be

$$i(S, n)$$

Then the difference d is

$$d = i(a, n) - i(S, n)$$

If this difference is assumed as quasi-constant, the short period-average $i(a, n)$ can be reduced to the normal period N for the whole of which only the station S has been operated. The reducing equation reads:

$$i(a, N) = i(S, N) + d$$

$i(a, N)$ is the average temperature at the place a reduced to the *normal period*.

The temperature series of the following table may serve as an example: It is assumed that the stations New York (Y), and New Haven (H) have been operated simultaneously only in the following January months:

	Y, °F	H, °F
January 1912	23 2	21.2
1914	31.1	29.2
1917	32 1	30 6
1918	21 3	19 8
1920	23.8	21.4
Average of 5 years	26 3	24.5

$$d = i(H, 5) - i(Y, 5) = -1.8 \text{ F}°$$

If New York is the normal station which has been operated during the entire 10-year period 1911 to 1920, we get the equation:

$$i(H, N) = i(Y, N) + d$$

and here

$$i(H, 1911 \text{ to } 1920) = i(Y, 1911 \text{ to } 1920) - 1.8 \text{ F}°$$

According to Table 35:

$$i(Y, 1911 \text{ to } 1920) = 31.0°\text{F}$$

Therefore the reduced value

$$i(H, 1911 \text{ to } 1920) = 29.2°F$$

In reality, the station at New Haven has also been operated over the entire 10-year period. The observed average is by chance identical with the calculated value.

It may be remarked that the 5-year average at New Haven is nearly 5 F° lower than the reduced one. Therefore the 5-year average cannot be compared with a 10-year average of another station.

The reduction to equal periods is fundamental for all climatological and climatographical purposes.

XII. 3. b. *The Reduction of Precipitation Series: Method of Quotients*

For series representing *rainfall*, the procedure of reduction to a normal period is analogous, but ratios are used instead of differences.

A normal (secular) station S has been operated over the normal period of N years. Another station a, located at a small distance from S, has been operated n years, so during these n years simultaneous observations have been conducted at the two stations.

The nomenclature is (partly repeated):

N = normal period of the Secular Station S

n = short period during which the station a has been operated $(n < N)$.

$P(S, N)$ = Normal amount of rainfall at S during the normal period N

$p(a, n)$ = average rainfall at a during the short period n

$P(S, n)$ = the same for the normal station

$p(a, N)$ = average annual rainfall at a, reduced to the normal period

$p(a, N)$ is the unknown value. Then under the assumption of quasi-constancy of the ratio q:

$$\frac{p(a, n)}{P(S, n)} = q = \frac{p(a, N)}{P(S, N)}$$

Therefore

$$p(a, N) = q \times P(S, N)$$

To take an example from Table 36A: Let Boston Weather Bureau (B) be the "normal" station, operated during the entire period

1911–1920.   At the Cambridge Waterworks, Waltham, Mass. (W), precipitation might have been measured for only five years, with the following results at the two stations:

July	B	W
1913	2 69	3 29
1914	2 64	2 46
1917 ..	1 10	0 93
1918	2 64	2 99
1920.	1 56	1 78
sum	10 63	11 45

$$p(W, 5 \text{ years}) = 11.45 \text{ inches}$$
$$P(B, 5 \text{ years}) = 10.63 \text{ inches}$$

and

$$\frac{p(W, 5 \text{ years})}{P(B, 5 \text{ years})} = \frac{11.45}{10.63} = 1.08 = \frac{p(W, 10 \text{ years})}{P(B, 10 \text{ years})}$$

According to Table 36, the 10-year average for Boston, Mass., is 3.96 inches, therefore: $p(W, 10 \text{ years}) = 1.08 \times 3.96$ inches $= 4.28$ inches.   As is seen from Table 36, the true 10-year average for Waltham is 4.10, so that the difference

$$\text{observed} - \text{calculated} = -0.18 \text{ inches}$$

The error is 4 39%.   It is a rather good agreement if one considers that the actual 5-year average for Waltham was 2.29 inches, against a true mean of 4.10 inches.

The reduction, reasonably applied, always yields reasonable results, while the comparison of different periods leads to false inferences.

XII.   3. c. *Physical Explanation of the Method of Quotients*

A short discussion must be added to explain why "ratios" instead of differences are used for the reduction of rainfall amounts.

There are two stations, a and b, with the average rainfall amounts $p(a)$ and $p(b)$.   The assumption is made that $p(b) > p(a)$. This may be expressed as follows· $p(b) = p(a) + \Delta p(a)$.

In a given year, a precipitation greater or smaller than normal may occur at a, that is, $k \times p(a)$, where k is a coefficient greater or smaller than 1.

The question is, what amount is recorded at b.

If the *differences* were invariable, $p'(b) = k \cdot p(a) + \Delta p(a)$.

Then $\Delta p(a)$ would be the constant difference between the amounts at the two places.

This assumption does not appear likely. The average increment $\Delta p(a)$ is caused at the place b, for example, by the contour of the region or some other physical feature   It is, therefore, much more probable that $p(a)$ and the increment $\Delta p(a)$ are increased according to the ratio k. We have therefore to expect at the place b the amount $kp(a) + k\Delta p(a)$.

Numerous examples which prove this consideration could be presented; hence it can be taken as certain that the *ratios* of the rainfall amounts at places not too far apart are quasi-constant, in contradistinction to their differences.

XII. 3. d. *The Reduction of Other Elements to a Given Period*

From the foregoing, it is clear that averages of temperature and precipitation at different places are comparable with one another only if they are related to an identical period. This postulate is obviously valid for each climatic element, be it primitive, derived, or combined.

There remains the question whether differences or quotients have to be chosen. Only a few of the most important elements are analyzed in this respect, namely:

a) Number of days with precipitation
b) Cloudiness [10]
c) Duration of snow-cover [11]
d) Number of days with fog

As the differences in the case of these various climatological elements yield better results than ratios, or results equally good, the method of differences is preferred because it involves easier calculating.

XII. 3. e. *Limits of the Method of Reducing Climatological Series to a Normal Period*

The basic principle of the "reduction" to a normal period is the quasi-constancy of differences or quotients. The mathematical

---

[10] V. Conrad, "Zum Studium der Bewölkung," *Met. Zeit*, 1927, pp. 87–91.
[11] V. Conrad, M. Winkler, "Beitrag zur Kenntnis der Schneedeckenverhältnisse in den österreichischen Alpenländern," *Gerland's Beiträge zur Geophysik*, XXXIV, 473–511.

meaning of the word "quasi-constancy" is that the variability of differences and quotients is much smaller than the variability of the elements themselves.   From the example in Table 35 may be derived:

$$v \text{ (diff.)} \quad = \pm 0.43$$
$$v \text{ (temp } Y) = \pm 4.89$$
$$v \text{ (temp. } H) = \pm 5.08$$

As was said earlier, v (diff.) is about a twelfth of the variability of temperature.

Obviously, the average variability of the differences (quotients) increases with increasing distance between the two places.   The same effect occurs if one of the two places is shifted beyond a climatic divide.   In both cases, the variability of the differences becomes greater and greater; continuously in the first case, and eventually discontinuously in the second.

Apparently, the quasi-constancy disappears when the variability of the differences (quotients) equals the variability of the element itself; then a reduction to a normal period is no longer reasonable.   That is the *theoretical* limit of the reduction, and is expressed by the equation.

$$v \text{ (diff., quot.)} = v \text{ (element)}$$

If v (diff. or quot.) is called $v(R)$ and v (element) is called $v(E)$ then generally this equation is valid:

$$v(R) = k \cdot v(E)$$

Only if $k < 1$ is a reduction theoretically admissible.

The question is, at what value of k should the *practical* limit of reduction be assumed?   Such limits must be determined empirically.   In the writer's experience, the reduction should be made only when

$$k \leqq 2/3$$

This result is, of course, valid for all elements.

The increase of $v(R)$ has been represented by Hann by means of an empirical, analytic equation, which reads:

$$v(R) = a + b \cdot \Delta E + c \ \Delta h$$

where a, b, c, are constants, ΔE the distance between two places, and Δh the difference of height.

This equation is mentioned here because it appears in many manuals, and also because it can be misleading.

At the limit:

$$\Delta E = 0 \qquad \Delta h = 0$$

the differences are related to the element of the "normal station" itself. Then, however, $v(R) = 0$.

This does not occur in the equation above, because the constant $a \neq 0$.[12]

In Chapter IV, 2, phenomena of decay are discussed and an example is given to show how the variation of $v(R)$ for cloudiness can be represented by an analytical equation which exactly fulfills the boundary conditions.

Lack of space prevents a fuller discussion of the problem; but the above hints should suffice for further investigations.

Reduction to a normal period has to be combined with an examination of the relative homogeneity, the equally important investigation into the nature of a climatological series. Both procedures require the computation of deviations.

Finally, it may be remarked that, in the opinion of many authors, very long series are comparable with one another without being reduced to the same period. Practically, one may be lucky enough to deal with series which happen to be comparable. It is perhaps more likely with non-reduced long series than with short ones.

In any case, however, especially with long series, the examination of relative homogeneity is unavoidable.

XII. 3. f. *Additional Note. The Length of a Normal Period*

It is frequently asked, how long a normal period should be. This depends upon three main conditions:

1) How many years of incontestable observations of the normal station or stations are available.
2) How long the whole period must be in order to include the periods of the series which have to be reduced.
3) How great the variability is of the element in question.

Because of these conditions it is not possible to define *a priori* how many years make a normal period.

[12] For more details see V. Conrad, "Zum Studium der Bewölkung," *Met. Zeit.*, 1927, p. 87.

Table 40 gives examples of the enormous variability of rainfall at Malden Island (4°S). In view of this fact, perhaps 50 years may not suffice to report, even approximately, the variations possible at this place. On the other hand, the variability of temperature on this tiny island in the equatorial belt of the Pacific Ocean, is nearly negligible. With regard to temperature, a normal period of 2 years would be sufficient.

Some scientists are of the opinion that the normal period should include at least one "Brückner-cycle," i.e., about 35 years; that is, a normal period should not be shorter than 35 years. The following considerations lead perhaps to the same result.

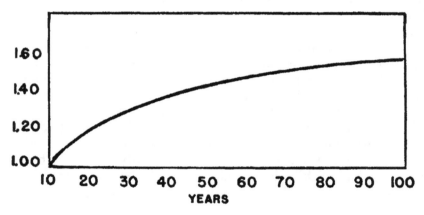

Fig. 27.   Increase of range of the extremes of temperature with length of period

Birkeland and Frogner [13] have shown how the range of the extremes of temperature and rainfall increases with an increasing period of observation. Their investigation indicates that this correlation (extreme range–length of period) can be computed from the Gaussian law of errors. An example is given for the range of the extremes of temperature in Fig. 27. The range of a period of 10 years is taken as the unit; this unit range is increased by 20% during a further 10-year period of observation.

The increase from 40 to 100 years is but about 20% as a consequence of the asymptotical trend of the curve. The increment from 100 to 150 years is 3%, and 2% for the following 50 years. These theoretical considerations are valuable hints, when making a reasonable choice of the length of a normal period, if a long series

[13] B J Birkeland, F. Frogner, "Die extreme Variabilität der Lufttemperatur," *Met Zeit.*, 1935, p. 349.

of observations is available at all.   But it is more important to take that portion of the long series which is incontestably homogeneous than to strive for a "normal period" as long as possible.

The upper limit of the length of a normal period should be dictated rather by the condition that the periods of the other series be included than by the intention of extending the normal period uncritically to infinity.   It is undesirable for obvious reasons that a normal period of 100 years be taken, if the longest of the other series do not cover more than perhaps 20 years.

These are some of the ideas which should be considered when choosing a "normal period."   In any case, its length should exceed 10 years, because otherwise the conclusions from statistical methods are no longer useful.   From all points of view, the length of a normal period should be about 25 to 35 years.[14]

XII.   3. g. *Interpolation of Missing Observations*

The quasi-constancy of differences or quotients, offers the possibility of interpolating missing or false values.   Curve parallels for the variations of the respective element at two and more stations are often useful, but numerical differences, and quotients in particular, should be preferred.   This is the more essential as these calculations must be made in order to investigate homogeneity, etc.   Systematic and random errors are revealed by these simple methods, and can often be corrected to a certain degree. Important corrections should be made—if possible—by means of at least two pairs of stations, each of which includes the dubious station.[15]

Only a full comparison of the synchronous climatological series used yields reliable results.

XII.   3. h.   *Coherent and Incoherent Climatic Regions and Their Statistical Evidence*

The increase of the variability of differences (quotients) with increasing distance between the places compared was mentioned in connection with the method of reducing climatological series to an identical period.   Although previously not emphasized, the

---

[14] The International Meteorological Organization adopted the period 1901–1930 as a "standard period" for climatological normals (Warsaw 1935).   See the *Meteorological Glossary* (London, 1939), under "Normals"

[15] For details, see V. Conrad, "Die klimatologischen Elemente," p. 119, chapter 7, The interpolation of missing observations.

increase of the variability of differences occurs more markedly if the difference in elevation between the places selected is increased. The increment of variability is a continuous process under otherwise equal conditions. This has been shown in a study of cloudiness.[16] As was said before, one may represent the increase of $v(R)$, the variability of the differences (quotients) with an exponential equation. The values calculated with this equation are in close agreement with observation Hann's formula, even if not in accordance with actual conditions in the immediate surroundings of the normal station, shows too that the differences increase continuously. Since this behavior holds for two climatological elements, one can assume that the same is true for other closely related elements.

This problem can be inverted in so far as the kind of variation of $v(R)$ can be taken as a climatological characteristic of a region.

A region within which the variability of differences (quotients) of elements at two places increases *continuously* with increasing distance (difference in elevation) is called *climatically coherent*. Lines or strips at which the variability of differences (quotients) varies discontinuously are called *climatic divides*. Therefore regions on different sides of a climatic divide are called *climatically incoherent*.

These concepts, derived from the variability of the differences (quotients) of elements at two places, are of major importance for recognizing the climatical structure of a region, and are fundamental for conclusions regarding the geographical distribution of the climatic elements Consequently, the short and long range forecaster is most concerned with these ideas A forecast fully valid for climatically coherent regions fails if it is extended to an adjacent, but climatically incoherent, region [17]

These considerations, coupled with those of homogeneity, lead to the method of correlation, most important in so many climatological problems, which will be discussed in the following intermediate chapter. The behavior of the correlation factor may be decisive whether or not climatic regions are coherent. This idea offers a great field for future investigations.

[16] V Conrad, "Zum Studium der Bewolkung," *Met Zeit*, 1927, pp 87 ff
[17] See also I I Schell, "On the Use of Climatically Coherent Areas in Seasonal Foreshadowing," BAMS, vol 23 (1942), pp. 182–183

INTERMEDIATE CHAPTER (C)
CORRELATION

C. 1. LINEAR CORRELATION

MATHEMATICALLY, the "quasi-constancy" of, for example, differences between average temperatures at two places means that a certain relationship exists between the temperature series at the two places. The ratio between the variability of the differences and the variability of the element itself is the measure of the association of the two temperature series.

This measure is specifically climatological and should be stated in terms of those measures offered by the statistical method of *correlation*. A generally valid measure of the degree of association of two series—in the simplest case—can be easily established.

Two series may consist of the following elements:

$$X_1, X_2, \cdots X_n$$

and

$$Y_1, Y_2, \cdots Y_n$$

The elements of these series have to be compared with one another in pairs, i.e., X_1 with Y_1; X_2 with Y_2; etc.

Because of the inequality of the arithmetical means ($\bar{X}$ and $\bar{Y}$), the direct comparison is possible only by introducing deviations:

$$X_i - \bar{X} = x_i \quad \text{and} \quad Y_i - \bar{Y} = y_i$$

There is no doubt that the *sum* of the products $x_i y_i$ is a measure of the degree of association of the two series.

If the number of the elements is great and there is no relationship between the two series, the products $x_i y_i$ are random numbers which may vary between certain definite limits. Then it must be possible to find for each product another of equal magnitude and opposite sign. (See the analogous conclusion with Abbe's criterion.)

This means that the sum of all products is zero if *no* association exists. Furthermore the value of $\sum_{i=1}^{i=n} x_i y_i$ depends, obviously, on the number n of the elements of the series. Therefore the sum of the products has to be divided by n, in order to free it of the influence of the arbitrary number of elements and we get

$$\frac{1}{n} \cdot \sum_{i=1}^{i=n} x_i y_i$$

There are two further difficulties in comparing the x_i series and the y_i series:

1) The different elements may be expressed by different units.
2) The scattering of the deviations of the two series is different.

If the foregoing expression is divided by the products of the standard deviations of the two series, it is reduced to equal scattering and the different units are eliminated at the same time. According to what has been said previously,

$$\sigma_x = \sqrt{\frac{\Sigma x_i^2}{n}} \qquad \text{and} \qquad \sigma_y = \sqrt{\frac{\Sigma y_i^2}{n}} \qquad \text{(See III. 2.)}$$

Therefore the measure of the degree of association of the two series is:

$$r_{xy} = \frac{\dfrac{1}{n} \sum_{i=1}^{i=n} x_i y_i}{\sqrt{\dfrac{\Sigma x_i^2}{n} \dfrac{\Sigma y_i^2}{n}}}$$

or finally

$$r_{xy} = \frac{\Sigma x_i y_i}{\sqrt{\Sigma x_i^2} \cdot \sqrt{\Sigma y_i^2}}$$

which appears often in the form

$$r_{xy} = \frac{1}{n} \frac{\Sigma x_i y_i}{\sigma_x \cdot \sigma_y}$$

The expression r_{xy} is called *correlation coefficient*. Obviously

$$r_{xy} = +1$$

if the functions $f(x_i)$ and $f(y_i)$ are identical or differ by a constant factor. In other words, the two series run parallel to each other.

If the trends of $f(x_i)$ and $f(y_i)$ are in exact opposition:

$$r_{xy} = -1$$

as it is clear from the equation for r_{xy}.

These are the boundary values that occur if the association of the series is perfect and can be expressed by means of a normal analytical equation. As stated before, $r_{xy} = 0$ if no association exists between the two series.

On the other hand, it is clear that a correlation coefficient greater or less than zero can often result by chance, even if no association between the two series exists, when only a small number of observations is available. In this case, it is unlikely that $\Sigma x_i y_i$ will become zero; consequently the correlation factor r differs from zero also for unrelated series. Even with a greater number of elements r will be falsified in a similar way.

It can be said, however, that then the true correlation coefficient r, which would result from an infinite number of elements of the two series, lies—with equal probability—within or without the limits $r + f$ and $r - f$. The value f is the probable error mentioned previously (III. 6). It is:

$$f = \pm 0.6745 \frac{1 - r^2_{xy}}{\sqrt{n}}$$

With an increasing number of elements the probable error decreases more and more, approaching zero. (See Table for $1/\sqrt{n}$ in Appendix II.)

The smaller f is, the more significant is the respective correlation coefficient. K. Pearson gives the rule of thumb that correlation coefficients should be assumed as real only if they are at least 6 times greater than the probable error. According to the author's experience, this ratio is very useful in avoiding false conclusions.

On the other hand, the limit $r_{xy} \geqq 6f$ is too high, according to other authors, for estimating the reality of correlation-factors.

In this connection, the opinion in the *Meteorological Glossary* (London, 1939, p. 54) is quoted:

Unless it is confirmed by physical reasoning or other independent evidence, a correlation coefficient should not be accepted as significant, unless it exceeds three times its probable error, in which case the odds in favor of significance are 20 to 1. If a number of trial correlations are made, the chance of obtaining a single large coefficient is obviously greatly increased, and such an isolated coefficient should not be accepted, unless it is four or five times its probable error.

In the main, it is of greatest importance to calculate not only the correlation factor but also the probable error. Only then is it possible to estimate the significance of a correlation factor.

Generally, correlation coefficients considerably smaller than 0.5 have to be estimated most critically.[1]

On the other hand, cartographical representation of correlation factors are very instructive. If a great region is covered by one sign and another region by the opposite sign and the two regions are systematically separated by a transition zone of mixed signs, in this case small correlation coefficients can also lead to further successful investigations.

If the correlation of a pair of series covering a great number of years shows a certain sign, the series should be divided into parts. For each of them, a correlation coefficient should be computed. If the signs of the single coefficients are identical, a certain significance even of small coefficients cannot be rejected without argument.

C. 2. EXAMPLE OF CALCULATING A CORRELATION

The average air pressures in millimeters of mercury (X_1) on the one hand, and the numbers of hours with bright sunshine (Y_1) on the other, are known for 10 months of February at a place in the temperate latitudes. $\bar{X}$ and $\bar{Y}$ are the arithmetical means, x_1 and y_1 the deviations. (Table 41.)

The opinion is frequently expressed that high pressure corresponds to fair, sunny weather in temperate latitudes. This assumption is examined by means of the values X_1, Y_1 in columns 2, 3 of Table 41. The correlation coefficient is below 1/2, and is not even double its probable error.

Therefore we have to conclude that a series of 10 pairs of average values does not offer any reason for assuming an association between air pressure and synchronous hours of bright sunshine, as far as averages (sums) and the region in question are concerned.

The plus sign ($+$) could indicate the tendency in the sense of the opinion mentioned; it would be significant only if it reappeared also with other equivalent pairs of series.

[1] Further remarks on this problem will be found at the end of C. 4. For an excellent model for this procedure, see Ellsworth Huntington, *Tree Growth and Climatic Interpretations in Quaternary Climates* (Washington Carnegie Institution, 1925), p. 161 ff. Attention is called also to the interesting "Dot Charts" of the relation between annual growth of sequoias and rainfall.

TABLE 41. EXAMPLE OF CALCULATING A CORRELATION

(1) No	(2) X_i	(3) Y_i	(4) x_i	(5) y_i	(6) $x_i y_i$	(7) x_i^2	(8) y_i^2
1	43 7	60	−3 0	−26	+ 78 0	9 0	676
2	51.5	133	+4.8	+47	+225.6	23 0	2209
3	50 6	72	+3.9	−14	− 54 6	15 2	196
4 .	47 9	82	+1 2	− 4	− 4 8	1 4	16
5	44.6	119	−2.1	+33	− 69 3	4 4	1089
6. .	47 2	52	+0 5	−34	− 17 0	0 3	1156
7	52.9	81	+6 2	− 5	− 31.0	38 4	25
8	41 0	54	−5 7	−32	+182 4	32 5	1024
9 ..	37 1	83	−9 6	− 3	+ 28 8	· 92.2	9
10	50 1	129	+3 4	+43	+146.2	11 6	1849

$$\bar{X} = 46\ 7 \qquad \bar{Y} = 86\ 5 \qquad\qquad \Sigma = +484\ 3 \qquad 228\ 0 \qquad 8249$$

$$r = \frac{\Sigma x_i y_i}{\sqrt{\Sigma x_i^2 \cdot \Sigma y_i^2}} = \frac{+484}{\sqrt{228 \times 8249}} = +0\ 353 \qquad\qquad r^2 = 0\ 125$$

$$1 - r^2 = 0\ 875$$

$$n = 10$$

$$f = \pm 0\ 6745 \frac{1 - r^2}{\sqrt{n}} = \pm 0\ 187 \qquad\qquad \sigma_x = \sqrt{\frac{\Sigma x_i^2}{n}} = 4\ 78$$

$$\frac{f}{r} = \frac{1}{2} \quad \text{or} \quad \frac{r}{f} = 1\ 9 \qquad\qquad \sigma_y = \sqrt{\frac{\Sigma y_i^2}{n}} = 28\ 72$$

Note: For calculating f see Appendix II ˙

To avoid a misunderstanding, it should be expressly remarked that the correlation factor does not indicate anything about the causal association of two series of numbers representing the variations of two geophysical phenomena. The indications of the correlation coefficients are only of *formal* nature. This fact cannot be sufficiently emphasized. Two series can show a high correlation coefficient, because of a systematic observational mistake, or because the association is physically well justified, as, for instance, for synchronous values of air pressure at two places a few miles apart. Therefore it is not sufficient merely to calculate the correlation factors. If they are known, the really scientific investigation begins with the physical explanation of the correlations computed.

In the example of Table 41, the explanation is obvious. The tendency toward the occurrence of precipitation is evidently smaller with high pressure than with low pressure. But above a plain, especially above a city, fog is much more frequent with high pressure conditions in winter. The final result is that a direct connection between air pressure and the number of hours with bright sunshine is lacking in this special case.

C. 3. SIMPLIFICATIONS IN COMPUTING CORRELATIONS

The number of applications of the method of correlation is unlimited in climatological problems; so some simplifications of the technique of calculation may appropriately be indicated here.

The computation of the deviations (x, y_i) is sometimes wearisome. If the simple definitions previously mentioned are considered, the following identities result:

$$x_i = X_i - \bar{X} \qquad \text{and} \qquad y_i = Y_i - \bar{Y}$$

$$\sum_1^n \bar{X} = n\bar{X} \qquad \text{and} \qquad \sum_1^n \bar{Y} = n\bar{Y}$$

$$\sum_1^n X_i = n\bar{X} \qquad \text{and} \qquad \sum_1^n Y_i = n\bar{Y}$$

Therefore

$$\sum_1^n x_i y_i = \sum_1^n X_i Y_i - n\bar{X}\bar{Y} \qquad \text{(see footnote}^2\text{)}$$

and

$$\sum_1^n x_i^2 = \sum_1^n (X_i - \bar{X})^2 = \sum_1^n X_i^2 - n\bar{X}^2$$

$$\sum_1^n y_i^2 = \sum_1^n (Y_i - \bar{Y})^2 = \sum_1^n Y_i^2 - n\bar{Y}^2$$

---

[2]

$$x_i y_i = (X_i - \bar{X})(Y_i - \bar{Y}) \quad \text{(Definition of } x_i \text{ and } y_i)$$
$$= X_i Y_i - Y_i \bar{X} - X_i \bar{Y} + \bar{X}\bar{Y}$$

Then

$$\sum_1^n x_i y_i = \sum_1^n X_i Y_i - \sum_1^n Y_i \bar{X} - \sum_1^n X_i \bar{Y} + \sum_1^n \bar{X}\bar{Y}$$

Because

$$\sum_1^n Y_i \bar{X} = \bar{X} \sum_1^n Y_i = n\bar{X}\bar{Y}$$

and

$$\sum_1^n X_i \bar{Y} = \bar{Y} \sum_1^n X_i = n\bar{X}\bar{Y}$$

and

$$\sum_1^n \bar{X}\bar{Y} = n\bar{X}\bar{Y}$$

we get:

$$\sum_1^n x_i y_i = \sum_1^n X_i Y_i - n\bar{X}\bar{Y} - n\bar{X}\bar{Y} + n\bar{X}\bar{Y} = \sum_1^n X_i Y_i - n\bar{X}\bar{Y}$$

The further transformations are analogous.

These expressions substituted in the original equation:

$$r_{xy} = \frac{\sum\limits_{1}^{n} x_i y_i}{\sqrt{\sum\limits_{1}^{n} x_i^2 \cdot \sum\limits_{1}^{n} y_i^2}}$$

yield

$$r_{xy} = \frac{\sum\limits_{1}^{n} X_i Y_i - n\bar{X}\bar{Y}}{\sqrt{\left(\sum\limits_{1}^{n} X_i^2 - n\bar{X}^2\right) \cdot \left(\sum\limits_{1}^{n} Y_i^2 - n\bar{Y}^2\right)}}$$

This formula uses only the quantities of the original series, avoiding the calculation of deviations.

If we look at the numbers in columns 2 and 3 of Table 41, the original items, a further trouble is apparent; the large numbers with which one has to calculate. This inconvenience also is easily avoidable, if similar, simple considerations are made as above. Each number of the two series is reduced by a number which equals the minimum of the respective series. That is 37.1 in the series of column 2, and 52 in the series of column 3. Thus we get small numbers for the calculation quickly and easily.

TABLE 42.   CALCULATING A CORRELATION BY MEANS OF THE
ORIGINAL NUMBERS REDUCED BY A CONSTANT VALUE

No.	X_i'	Y_i'	$X_i'Y_i'$	$X_i'^2$	$Y_i'^2$
1.. ...	6 6	8	52.8	43.6	64
2 .	14 4	81	1166 4	207 4	6561
3 . .	13 5	20	270.0	182 2	400
4 . . 10.8		30	324 0	116.6	900
5 ...	7.5	67	502.5	56 3	4489
6. ... .10.1		00	0 0	102 0	0
7. .	15 8	29	458.2	249.6	841
8 .	3 9	2	7 8	15 2	4
9.	0 0	31	0.0	0 0	961
10.	13 0	77	1001 0	169.0	5929

$\bar{X}' = 9\,6$ $\bar{Y}' = 34.5$ $\Sigma X_i'Y_i' = 3782\,7$ $\Sigma X_i'^2 = 1141.9$ $\Sigma Y_i'^2 = 20149$

$n = 10$; $n\bar{X}'\bar{Y}' = 3312.0$; $n\bar{X}_i'^2 = 921\,6$, $n\bar{Y}_i'^2 = 11902\,5$

$$r_{xy} = \frac{\Sigma X'Y_i' - n\bar{X}'\bar{Y}'}{\sqrt{(\Sigma X_i'^2 - n\bar{X}'^2)(\Sigma Y_i'^2 - n\bar{Y}'^2)}} = \frac{470\,7}{1347\,9} = +0\,349$$

The example of Table 42 is identical with that of Table 41. The correlation between pressure (X_i) and sunshine hours (Y_i) is calculated with these original numbers instead of the deviations x_i and y_i. Furthermore, X_i and Y_i are reduced by the minimum values of the respective series.

The new nomenclature is:

$$X_i - 37.1 = X_i' \qquad \text{and} \qquad Y_i - 52 = Y_i'$$

The other details of Table 42 are understandable without explanation. The results of the calculations in Tables 41 and 42 are, practically, identical.

C. 4. Regression Equation

The deviations from the arithmetical mean x_i and y_i (see C. 1.) can be plotted in pairs on a graph paper representing a Cartesian system of coördinates with the origin $x = 0$ and $y = 0$. The points x_iy_i can be scattered in such a way that they cover the quadrants of the coördinates with the greater uniformity the greater the number n of the pairs of deviations is. This is not of interest here. On the other hand the dot-diagram can show a crowding of the points, indicating a linear proportionality in the simplest. (See Fig. 28.) Then we can try to determine a straight line $y = bx$ which passes the origin and is fitted to the point-cloud as well as possible.

If x should be an analytical linear function of y then $y - bx$ would be exactly zero. In this problem, however, this is not valid, and the quantity b must be determined so that it makes the difference $y - bx$ as small as possible.[3] Since the signs of the deviations

---

[3] The reader may profit by another very expedient method of explaining, kindly communicated to the author by L. W. Pollak, Dublin, Ireland, in a letter. I quote.

"We have observed n pairs of phenomena X_i, Y_i. Forming the means of X_i and Y_i respectively, we obtain

$$X = \bar{X}_i = \frac{1}{n} \Sigma X_i, \qquad Y = \bar{Y}_i = \frac{1}{n} \Sigma Y_i, \tag{1}$$

The values X, Y define in the graphical representation of our X_i, Y_i a point M which as the 'centre' of the dot diagram plays an important role.

"Further, we compute the deviations of the X_i and Y_i from their arithmetical means, obtaining

$$x_i = X_i - \bar{X}_i = X_i - X, \qquad y_i = Y_i - \bar{Y}_i = Y_i - Y \tag{2}$$

Plotting the x_i, y_i means nothing but shifting the origin of the coördinate system of the X_i, Y_i dot diagram to M.

"Now we intend to approximate our deviations x_i, y_i (in the coördinate system with

do not play any role in this problem, only their absolute amounts, b should be chosen so that the sum of the *squares* of $y - bx$ is as small as possible. This is reached if $\Sigma(y - bx)^2$ becomes a minimum. The condition is fulfilled, if

$$\frac{d}{db} \Sigma(y - bx)^2 = 0$$

and

$$2\Sigma(-x)(y - bx) = 0 \qquad \text{or} \qquad \Sigma xy = b\Sigma x^2$$

Therefore

$$b = \frac{\Sigma(xy)}{\Sigma x^2} = \frac{\frac{1}{n}\Sigma(xy)}{\frac{1}{n}\Sigma x^2}$$

The equation $y = bx$ contains the arbitrary assumption that x is the independent variable.

One can also write:

$$x = b'y$$

and

$$b' = \frac{1/n \cdot \Sigma(xy)}{1/n \ \Sigma y^2}$$

M as origin) by a straight line

$$y = bx + d \tag{3}$$

so that the sum of the squares of the distances of each dot from our straight line, measured in the direction of the y's, is a minimum.

"Therefore

$$\Sigma[y_i - (bx_i + d)]^2 = \text{Min.} \tag{4}$$

and accordingly

$$\left.\begin{array}{l} \dfrac{\partial}{\partial b} \Sigma[y_i - (bx_i + d)]^2 = 0 \\[2mm] \dfrac{\partial}{\partial d} \Sigma[y_i - (bx_i + d)]^2 = 0 \end{array}\right\} \tag{5}$$

or

$$\left.\begin{array}{l} 2\Sigma[y_i - (bx_i + d)](-x_i) = 0 \\ 2\Sigma[y_i - (bx_i + d)](-1) = 0 \end{array}\right\} \tag{6}$$

The second equation of (6) gives directly:

$$\Sigma y_i - b\Sigma x_i = nd$$

and

$$d = 0$$

as $\Sigma x_i = 0$ and $\Sigma y_i = 0$. (The sum of the deviations from the arithmetical mean, *regarding* the signs, is zero) Our straight line required passes through the origin or the 'centre' M."

If we consider that the standard errors

$$\sigma_x = \sqrt{\frac{\Sigma x^2}{n}} \qquad \sigma_y = \sqrt{\frac{\Sigma y^2}{n}}$$

and the correlation coefficient

$$r = \frac{\Sigma(xy)}{\sqrt{\Sigma x^2 \cdot \Sigma y^2}} = \frac{1}{n}\frac{\Sigma(xy)}{\sigma_x \cdot \sigma_y}$$

TABLE 43   ANNUAL VARIATIONS (DEVIATIONS) OF THE LEVEL OF VICTORIA NYANZA
(0°N, 32°E), (x, inches) and of the Annual Sunspot Numbers (y_i)

(*After Sir Napier Shaw*)

(1) Year	(2) x_i	(3) y_i	(4) x_i^2	(5) y_i^2	(6) $x_i y_i$
1902 ..	−18	−35	324	1225	630
1903	5	−16	25	256	−80
1904	10	2	100	4	20
1905	7	23	49	529	161
1906	21	14	441	196	294
1907	13	22	169	484	286
1908	2	9	4	81	18
1909	0	4	0	16	0
1910	− 7	−21	49	441	147
1911	−15	−34	225	1156	510
1912	−19	−36	361	1296	684
1913	−11	−39	121	1521	429
1914 .	−10	−30	100	900	300
1915	− 4	7	16	49	−28
1916	7	17	49	289	119
1917	27	64	729	4096	1728
1918	19	41	361	1681	779
1919	0	24	0	576	0
1920	− 5	− 2	25	4	10
1921	−13	−15	169	225	195
Σ			3317	15025	6202
mean			166	751	310

$$r_{xy} = \frac{1}{n}\frac{\Sigma xy}{\sigma_x \sigma_y} = +0.88 \qquad\qquad \sigma_x = \sqrt{166} = 12.9$$

$$b_{xy} = r\frac{\sigma_x}{\sigma_y} = 0.414 \qquad \underline{x = 0.414y} \qquad \sigma_y = \sqrt{751} = 27.4$$

$$b_{yx} = r\frac{\sigma_y}{\sigma_x} = 1.869 \qquad \underline{\underline{y = 1.869x}}$$

then

$$b = r \frac{\sigma_y}{\sigma_x} \qquad \text{and} \qquad b' = r \frac{\sigma_x}{\sigma_y}$$

Therefore we get:

$$y = r \frac{\sigma_y}{\sigma_x} x \qquad \text{and} \qquad x = r \frac{\sigma_x}{\sigma_y} y$$

These equations have been called *regression equations* by Sir Francis Galton (1886); $b(b')$ is called the *regression coefficient.*

An example (Table 43) [4] serves as an illustration.

Let x, denote the deviations (inches) from the average level of Lake Victoria (Victoria Nyanza) in Africa, and let y, denote the deviations from the mean of the annual sunspot numbers (period 1902–1920). The values of columns 2 and 3 in Table 43, x, and y, are plotted on graph paper. The values x, mean the variations of the lake level (horizontal axis) and the y, the variations of the sunspot numbers (vertical axis). This dot diagram (Fig. 28) shows clearly that there is a rather strong correlation between the two phenomena.

The two regression equations mentioned above are represented by two different straight lines in Fig. 28. The smaller the angle between the two regression lines, the higher the degree of association and the greater the value of the correlation-coefficient r. If $r = \pm 1$, the two regression lines coincide. If $r = 0$ the one line is identical with the X-axis and the other with the Y-axis ($y = 0$; $x = 0$).

As seen from Table 43, the correlation coefficient is high, and from Figure 28 it is clear that the angle between the regression lines is small in consequence of the strong association. Such cases are very rare in the realm of meteorology and climatology. The method of the curve-parallels is often misleading, and dot-charts are by far preferable to this method. Calculating correlation-coefficients and probable errors is never wasted time if the dot-chart is somewhat encouraging.

The correlation coefficient, however, and the probable error, do not yet entirely satisfy the climatologist (as we have mentioned before). The critical estimation of the true value of a correlation coefficient is difficult. W. H. Dines says: [5]

[4] Taken from C. E P Brooks, *Climate through the Ages* (New York, 1926), p 415 ff, and Sir Napier Shaw, *Manual of Meteorology*, I, 284, Fig. 116 and table on the same page The continuation of the example concerning the partial correlation is also taken from C E P Brooks' discussion of the problem

[5] *The Computor's Handbook* (Met Office, London, 1915), p V 39.

No precise classification of correlation coefficients can be made. Assuming that they depend on as many as 50 independent cases, one may say roughly that values under 0.30 are hardly significant, values between 0.30 and .70, prove a moderate connection; values between .70 and .90, a close connection, and values over .90, a very intimate connection indeed.

It is also apparent that, if the coefficient is very high, fewer observations are necessary to establish it, since $1 - r^2$ is then small.[6]

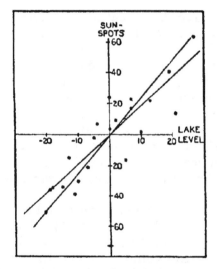

FIG 28   Dot-chart and regression lines of the deviations from mean annual level of Victoria Nyanza [0° (Eq), 32°E] and from the mean number of sunspots   The mean level of the lake and the mean number of spots are taken as axes.   (After Sir Napier Shaw)

From this classification, the coefficient of the example in Table 43 is only "close," and C. E. P. Brooks expresses this fact by saying that "there is a good, but by no means perfect, agreement between the variations of the level of Victoria Nyanza and sunspot numbers." We can assume that "variations of level are 'caused' partly by variations of sunspots and partly by other factors, which for the sake of illustration we shall suppose to be independent of sunspots." This portion of the variations of level is, however, $1 - r^2 = 1 - (0 88)^2 = 0.23$, or 23%.

C. 5. PARTIAL CORRELATION

In the foregoing, the problem of two variants which show a certain degree of association has been discussed. This case,

[6] See probable error.

apparently, is the most important for the climatologist. The general problem, however, is that there are m variants, mutually dependent upon one another. The question is, how and with what accuracy can x be represented by the other variants, y, z, $\cdots$.

The method starts from the aforementioned correlation coefficients and analyzes the complex influence of each of the series of variables upon x, so that the partial influence of the single variables upon x is separated one from the other. Therefore we speak of "partial correlation."

If, for instance, three variants are associated one with another, the problem is to find how the second variable is correlated with the first, when the third is constant, and so forth.[7]

Now we return to the example of Table 43, which indicates that the variations of the level of Victoria Nyanza (x) are closely associated with the synchronous variations of the numbers of sunspots (y).

In this example, especially, the need of a connecting link in the form of a third variant is evident. this is the rainfall. Its variations (z) are known, and the correlation between z and x is indeed high:

$$r_{xz} = +0.92$$

The coefficient between rainfall variations (z) and the variations of sunspot numbers (y) is:

$$r_{yz} = +0.80$$

Theoretically two alternatives exist:

1) The sunspots influence the rainfall, and its variations cause the variations of the level of the lake.
2) There is a direct connection between sunspots and lake level.

One should discriminate between these two alternatives. For this purpose, the partial correlation coefficients are calculated.

We repeat the nomenclature. if

r_{xy} = the coefficient between lake level (x) and sunspots (y)
r_{xz} = the coefficient between lake level (x) and rainfall (z)
r_{yz} = the coefficient between rainfall (z) and sunspots (y)

[7] C. U. Yule, *An Introduction to the Theory of Statistics* (2nd ed., London, 1912) and F. M. Exner, *Ueber die Korrelations Methode* (Jena, 1913).

the partial correlation coefficient [8]

$$r_{xy.z}$$

is given by the equation

$$r_{xy.z} = \frac{r_{xy} - r_{xz} \cdot r_{yz}}{\sqrt{(1 - r^2_{xz})(1 - r^2_{yz})}}$$

From the example we get:

$$r_{xy.z} = \frac{0.88 - 0.92 \; 0 \; 80}{0.39 \cdot 0.60} = + 0 \; 61$$

The meaning of $r_{xy.z}$ is that "the effect exerted through rainfall" is "eliminated." In other words, the correlation between x and y is examined at constant rainfall. The relatively small correlation coefficient $r_{xy.z}$ shows that the major part of the effect on the level of the lake is caused by rainfall; that corresponds well to common sense. On the other hand, "there is still an appreciable effect due to some cause independent of rainfall, most probably rate of evaporation."

The expression for $r_{xy.z}$ has been given. The two other coefficients are easily derived from this equation. We exchange y for z, z for y, and get from

$$r_{xy.z} \qquad r_{xz.y}$$

and an exchange of x for y, which in turn yields,

$$r_{yz.x} \qquad \text{from} \qquad r_{xz.y}$$

---

[8] D Brunt, *The Combination of Observations*, (Cambridge, England, 1931), p. 175, gives an interpretation of the partial correlation coefficient which may be quoted here as a very valuable addition to the above-mentioned explanation For the convenience of the reader, Brunt's symbols are replaced by those used here in the text I quote:

"We further require to know the correlation between x and y when each has been corrected for the correlation with z This is written $r_{xy.z}$, indicating that the effect of the variable z is eliminated. Since the regression equations of x and y on z are

$$x - r_{xz}\frac{\sigma_x}{\sigma_z} z = 0, \qquad y - r_{yz}\frac{\sigma_y}{\sigma_z} z = 0$$

$r_{xy.z}$ is the coefficient of correlation between

$$x - r_{xy}\frac{\sigma_x}{\sigma_z} z \qquad \text{and} \qquad y - r_{yz}\frac{\sigma_y}{\sigma_z} z."$$

Thus we get three partial correlation coefficients:

$$r_{xy.z} = \frac{r_{xy} - r_{xz}\,r_{yz}}{\sqrt{(1 - r^2_{xz})(1 - r^2_{yz})}}$$

$$r_{xz.y} = \frac{r_{xz} - r_{xy}\cdot r_{yz}}{\sqrt{(1 - r^2_{xy})(1 - r^2_{yz})}}$$

$$r_{yz.x} = \frac{r_{yz} - r_{xy}\cdot r_{xz}}{\sqrt{(1 - r^2_{xy})(1 - r^2_{xz})}}$$

The respective standard deviations are:

$$\sigma_{x.yz} = \sigma_x\sqrt{1 - r^2_{xy}}\cdot\sqrt{1 - r^2_{xz.y}}$$

$$\sigma_{y.xz} = \sigma_y\sqrt{1 - r^2_{xy}}\,\sqrt{1 - r^2_{yz.x}}$$

$$\sigma_{z.xy} = \sigma_z\sqrt{1 - r^2_{xz}}\cdot\sqrt{1 - r^2_{yz.x}}$$

The following six regression coefficients correspond to the three partial correlation coefficients.

$$b_{xy.z} = r_{xy.z}\frac{\sigma_{z.yz}}{\sigma_{y.xz}} \qquad b_{xz.y} = r_{xz.y}\cdot\frac{\sigma_{x.yz}}{\sigma_{z.xy}}$$

$$b_{yx.z} = r_{xy.z}\frac{\sigma_{y.xz}}{\sigma_{x.yz}} \qquad b_{yz.x} = r_{yz.x}\cdot\frac{\sigma_{y.xz}}{\sigma_{z.xy}}$$

$$b_{zx.y} = r_{zx.y}\frac{\sigma_{z.xy}}{\sigma_{x.yz}} \qquad b_{zy.x} = r_{yz.x}\cdot\frac{\sigma_{z.xy}}{\sigma_{y.xz}}$$

Finally, the regression equations read.

$$x = b_{xy.z}\cdot y + b_{xz.y}\,z$$

$$y = b_{yx.z}\cdot x + b_{yz.x}\cdot z$$

$$z = b_{zx.y}\,x + b_{yz.x}\cdot y$$

The example mentioned above shows that the method of the partial correlation is well suited for discriminating between the individual influences of the different variables. Therefore the method of partial correlation is much more than a kind of representation. it can be a successful method of investigation. For want of space, these remarks about linear correlations must suffice. Generally, the common climatological problems can be solved by the methods indicated.

L. W. Pollak's application of "autocorrelation" [9] is of great importance for determining hidden periodicities as well as the most significant problems of long-range forecasts as far as higher statistics are concerned. This method correlates two portions of the series with the series itself, which may briefly explain term and aim of the method.

[9] L. W. Pollak, "On Cycles of Pressure Especially in the Neighborhood of Symmetry Points," QJRMS, vol LXVI, no 287 (1940)

CHAPTER XIII

GRAPHIC COMPARISON OF CLIMATOLOGICAL ELEMENTS
ISOGRAM. MAP OF ISOLINES

THE PROBLEM is to represent three variables on a two-dimensional plane. A well-known solution is the representation of contour lines on a map. The first variable is the geographical longitude, plotted on the abscissa. The second variable, the latitude on the ordinate. Points of equal elevation (the third variable) are connected with one another by the contour lines. This analogy leads to the understanding of the method of repre-

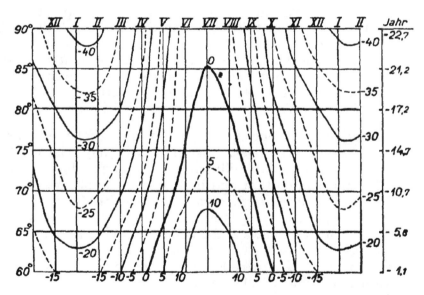

FIG 29   Isopleths representing the annual course of air temperature in the high northern latitudes. (After H Mohn and W. Meinardus)

sentation. An example of the application to climatological purposes is given in Fig. 29. The months are plotted on the X-axis, and the geographical latitudes on the Y-axis. The monthly temperatures of the parallels may be known. For 85°N, e.g., the

166

numbers read:

	Jan.	Feb.	Mar.	Apr.	May	Jn	Jl	Aug	Sept	Oct.	Nov.	Dec.
°C	−37	−37	−32½	−25	−12	−2	0	−2	−11	−22	−29	−34

These temperatures are plotted on the map at the intersections of the verticals (indicating the mid-points of the months) and the parallels. Now the points of equal temperature are connected by curves as in Fig. 29. Along these lines, the temperature is constant. They are called *isotherms* for temperature; (*isohyets* for precipitation, *isonephs* for cloudiness, *isohels* for sunshine, etc.)

The specific representation of three variants—of which, e.g., the first is the characteristic of an element, the second the time, and the third a linear dimension—is called *isopleth-diagram* or *isogram*.

Figure 29 takes the place of a table of 7 rows and 12 columns; besides,

1) by interpolation the annual course of temperature can be taken from this representation for any latitude between 60°N and 90°N.

2) The vertical lines mean the middle of the months. Thus, by interpolation, the temperature-decrease from 60°N to the pole can be read off along the vertical for any date of the year.[1]

On the other hand, one should bear in mind that the accuracy of interpolation is very poor. Be that as it may, the isopleth-diagrams are often valuable illustrations of climatological conditions. The most important and indispensable application of this general method is the *map of isolines*. Isolines are fully analogous to contour lines, or to the equipotential lines used in physics.

From this analogy, laws and rules for drawing isolines can be easily deduced Closed isolines which fully surround a region indicate that this is either depressed or elevated Isolines of equal value, running parallel, border an elevated ridge or a depressed furrow, etc. These principles, translated from the

[1] Attention should be drawn to a special kind of isograms, which show the daily course as well as the annual Irving F Hand, in "A Summary of Total Solar and Sky Radiation Measurements in the United States," MWR, vol 69 (April 1941), pp 95–125, publishes a great collection of this kind of isograms, which present the best pattern for such graphs They are extremely clear and legible, because they are not overburdened. The hours of the day are plotted on the X-axis, the months on the Y-axis

L. W. Pollak gave an original application of the method of isograms, constructing *iso-stereograms* which are certainly of great general value and are also useful for didactic purposes (L. W. Pollak, *Plastische Tabellen* [Plastic Tables], Prague, 1914)

language of the contour-lines into that of climatology, mean for instance: high and low pressure region, ridge of high pressure, trough of low pressure, saddle between two high pressure regions, and similar characteristics for each element which is represented by a map of isolines.

Isolines representing different values cannot cross one another. If the trend of an isoline is followed in a given sense, clockwise or counter-clockwise, decreasing and increasing intensities (amounts) always remain on the same side.   Every map of isolines should be strictly examined, according to this principle.[2]   Isolines should be drawn so that the difference of the intensity (amount) is constant between any pair of adjacent lines of the same map.   Isolines which deviate from this general rule should be especially marked (e.g. by broken lines, if the regular are full).

The mutual distance of two isolines is inversely porportional to the degree of variation (gradient) of the element.   The "density" of the isolines is proportional to the intensity of the spatial variation of the element.

The *linear* interpolation of the intensity (amount) of an element between two points (places), is a necessary expedient in default of a better one.   But the climatologist should realize how rough this approximation is.   This circumstance on the one hand, and the mass of errors with which the average values of the elements are infected on the other, should warn everybody against an exaggerated accuracy in planning and drawing such maps.   It must also be emphasized that climatological networks usually have wide meshes.   Thus, many lines are drawn over a distance of hundreds of miles, with only two or three points of known values.   It is regrettable that, sometimes, neither is the homogeneity examined nor a reduction to an equal period made.   These omissions have not deterred students from drawing isotherms for every F°, in spite of the fact that the actual exactness of these observations is about ±2 to 3 F°.   Consequently, the difference of intensity (amount) between two consecutive isolines should be weighed according to the accuracy of the data at hand, and to the openness of the network of stations.

The map of isolines must never be overburdened.   Therefore the use of interpolated isolines between those based on observations in order to give a better appearance to the map is strictly ruled out.   Such inserted parallel lines are nothing but an un-

[2] See also the short instruction in A F. Spilhaus and J. E. Miller, *Workbook in Meteorology* (New York, 1942), p 1.

necessary and arbitrary interpolation. Finally, the trend of the isolines should be chosen as simply as observations permit. Fancifully embellished trends mislead the layman, and are little appreciated by the expert.

As far as the projection of maps is concerned, Sir Napier Shaw and V. Bjerknes arrive at the same conclusions Shaw recommends in his *Manual* (vol. I, p. 262): ". . . three conformal projections, two polar planes extending to $67\frac{1}{2}°$, an adjusted conical projection for the region between latitude $67\frac{1}{2}°$ and $22\frac{1}{2}°$, and for the equatorial regions between $22\frac{1}{2}°N$ and $22\frac{1}{2}°S$, projection on a cylinder adjusted to be conformal."

The number of applications of the method of isopleths and isograms can be increased arbitrarily, like the number of derived elements. Therefore, in the following section and in Part IV, only a few hints can be given for making maps of isolines and some difficulties mentioned.

CHAPTER XIV

METHODS OF ANOMALIES

XIV. 1. Temperature Distributions in Mountainous Regions

TEMPERATURE varies with latitude according. to radiation conditions and the length of the day; it varies also with the distribution of land and water. Moreover, it varies with altitude about one thousand times as rapidly as with latitude. That is the difficulty. It is fortunate in this respect that about three fourths of the surface of the continents does not exceed 1300 feet, and that 58% is below 660 feet. In relation to these numbers, the mountain ranges appear only as disturbances of the continental surfaces, as far as the whole surface of the earth is concerned, or at least great portions of it. Therefore, the variations of temperature with altitude have little influence on the trend of the isotherms, on the whole.

To describe the climate of a mountainous country is another matter; for here the differences in altitude and relief play a decisive role, and the climate is determined by these factors.

XIV. 2 Examination of Lapse Rate

The variation of temperature per unit of height is called *lapse rate*. It is positive if the temperature decreases with height and is negative when it increases. The latter phenomenon is usually called *temperature inversion*.

Two methods of calculating the lapse rate are used.

1) The temperature difference of two places at different altitudes is computed, and divided by the difference of height. The single lapse rates are averaged.
2) As described in an earlier chapter, temperatures of the different stations are arranged according to their respective altitudes, then the elevations and the corresponding temperatures are averaged within certain intervals of altitude. (See IV, 4. b.) Thus a temperature-altitude curve is obtained, from which average lapse rates can be

taken for each interval, as well as for the entire difference of height. This latter method seems to be better and more accurate. The advantage of the first method is that a very small number of pairs of stations yield a first approximation.[1]

XIV. 3. Mapping Temperature Conditions of a Mountainous Country (Reduction to a Given Level)

The most common means of representation is by *reduction to a given level*. J. Hann suggested reducing temperatures in a uniform manner, independent of season and region, under the assumption of a lapse rate of 0.5 C° per 100 meters or 3 F° per 1100 feet. This suggestion should be generally accepted. Although reduced isotherms are not satisfactory in a country with great "relief energy"[2] and great differences in elevation, at least the comparableness of such isotherm maps would result. It is regrettable that uniformity is still lacking. Following Hann's rule, the temperature reduced to sea level, t_o, is computed by the formula

$$t_o = (t + 0.005\,h)°C$$

if t is the average temperature in °C at an altitude of h meters; or

$$t_o = (t + 0.002743\,h)°F = (t + 3/1100 \cdot h)°F$$

if h is expressed in feet.

Frequently, sea level is not the most advantageous plane, but any other level, e g. the average level of the valley bottom, or that of the plain or highland adjacent to the mountains, may often yield more plausible temperatures.

The reduction of temperature when there are great differences in altitude is no longer a correction but a method of projection onto a plane. This projection should offer:

1) A survey of the geographical distribution of the temperature.
2) The possibility of interpolating the temperature at a place of known altitude from which no observations are available.

---

[1] For numerical comparison between the two methods, see J Maurer, *Das Klima der Schweiz* (Frauenfeld, 1909), p 154.

[2] "Relief energy" is the sum of the differences in height along a cross section through the region, divided by the length of the cross sections. Obviously, this ratio is zero for a plain. Many authors relate the "relief energy" to a certain area and not to a length. (H. Wagner, W Meinardus, *Mathemat. Geogr.* 1938, p 349).

At first, this map may be disappointing because it contains fictitious temperatures which give a false picture. Only a conversion of the temperatures by means of the altitudes (which are mostly lacking on the maps) gives the actual temperatures.

If the picture presented by the reduced isotherms is unsatisfactory, what of the possibility of interpolation? Here, too, the method does not give good results for winter in temperate latitudes. The reason is simple, since temperature inversions are more or less common above the valley bottom. This means that the average lapse rate is negative, and the average temperatures increase with height. It is cold in the valley bottom and much warmer at higher levels. The consequences of the reduction of temperature to a certain level under these conditions are illustrated by the following example:

The temperature at a place A at the bottom of a valley at 440 m reduced to sea level is $-4.2°C$. The actual temperature of a place B, 746 m high, located on the slope, should be interpolated from the map. Then

$$t = t_o - 7.46 \times 0.5 \text{ C°} = -4.2 - 3.7 = -7.9°C$$

In reality, observations at B are at hand and the true January temperature is $-3.7°C$, which is about 4 C° higher than that interpolated from the isotherm map. The explanation is simple. We have calculated with a uniform lapse rate of 0.5 C°/100 m, whereas in reality there is an increase of temperature, which cannot be considered in the procedure of a uniform interpolation.

Consequently, one must conclude that the reduced isotherms do not yield a suitable representation of the temperature distribution in a mountainous region.

This conclusion, it should be emphasized, applies only in the case of a mountainous region. In open land, also in hilly regions, where small differences in height are more or less negligible in comparison with the respective horizontal distance, the reduction to a fixed level may be accomplished with good results.

XIV. 4. ACTUAL TEMPERATURES

Owing to the great disadvantages of the method of reduced temperatures, many authors are led to give "actual temperatures," i.e., the true average temperatures as observed at the altitudes of the respective places.

Two alternatives exist:

1) There is a huge country with a high mountain range which is relatively not extensive in area, and the number of mountain stations is small (United States).
2) There is a country the greatest portion of which is covered with high mountains (Switzerland).

In problem 1, the actual temperatures at the elevated stations do not affect the general picture of isolines much, since the mountain stations are few.   In the open, hilly regions, the representation of the temperature distribution differs but little from the reduced isotherms, because of the small differences in height.

In problem 2, the actual isotherms could be drawn only if the density of the network were exceedingly great.   Otherwise the curves would cross crests and abysses, valleys and ridges, regardless of the variations of temperature.   Yet just these variations should be represented.

A clearly legible number, giving the actual average temperature, beside each place-name (similar to the method of indicating the altitude on orographic maps), would be a much better expedient than lines, the trend of which is by no means certain and cannot be guaranteed, since the number of stations is never sufficient, owing to the great "relief energy."

It is generally recommended that numbers characterizing the represented elements be added beside the station names (or station symbols) on maps of isolines, in any case   Only then is the reader able to control the trend of the lines and to change it according to his individual point of view.

XIV. 5. STANDARD CURVES. ANOMALIES. ISANOMALS

As already discussed, the influence of altitude upon the geographical distribution of temperature should be eliminated by a sort of projection called *reduction to a fixed level*.   Another attempt in this direction, utilizing the *actual temperatures*, has been made. It has been shown in the foregoing that neither the one nor the other method yields satisfactory results.

The discussion in Chapter IV. 4, about "curve fitting," concluded that the graphical Cartesian representation of two variables, an independent (e.g., altitude) and a dependent (e.g., temperature), is important for climatological purposes.

These dot charts are interpolated either graphically or by means of analytical equations. Both methods permit the determination of the values of the dependent variant for every value of the independent. Thus convenient tables can be constructed from which the values of the dependent variant can be taken for equidistant values of the independent variant. A clipping from a table representing the average temperature-height curve above the bottom of a mountain valley was given in Table 16, as well as the corresponding graph Fig. 14, which can be called the *standard curve*. The numbers in the table represent the *standard distribution* of the element with respect to the region in question.

On the other hand, the actual value of the element at the place is known, as also its elevation. The difference, viz., actual value of the element minus standard value, is called *anomaly*.

These anomalies are independent of altitude in the present case, because this is fully considered by means of the standard curve. Lines connecting places of equal anomaly are called *isanomals*.

XIV. 6. APPLICATIONS OF THE METHOD OF ANOMALIES AND ISANOMALS

XIV. 6. a. *Vegetative Period and Altitude*

The length of the vegetative period in Switzerland may serve as the first example.[3]

As was said previously, the duration (number of days) of a temperature above 43°F, derived from an average annual course of temperature, is called the vegetative period.

These periods have been calculated by a simple interpolation formula (see VI. 2) for 168 stations in Switzerland. Their altitudes vary between about 800 and 8200 feet. The entire country, with mountains up to nearly 16,000 feet, has an east-west extension of about 186 miles, and a north-south extension of about 124 miles. This is, indeed, a mountainous country divided into a northern and a southern part by the crest-line of the Central Alps.

From all the 168 stations together, suitably grouped according to elevation, the following analytical equation for the relation between the duration in days (d) of the vegetative period and the

---

[3] V Conrad, "Isanomalen der Andauer einer bestimmt vorgegebenen Temperatur," *Geografiska Annaler*, 1929, p. 299.

J. Maurer (*Das Klima der Schweiz*, Frauenfeld, 1909, p. 68 ff) gave anomalies and isanomals of annual average temperatures in Switzerland.

altitude in meters (h) is obtained:

$$d = 268 - 0.07\,h$$

From this standard equation, the standard values are derived for altitudes for every 200 meters (Table 44).

TABLE 44. STANDARD VALUES (DAYS) OF THE DURATION OF THE VEGETATIVE PERIOD IN SWITZERLAND

Height (meters)	Duration (days)	Height (meters)	Duration (days)
200	254	1200	184
400	240	1400	170
600	226	1600	156
800	212	1800	142
1000	198	2000	128

The vegetative periods of two places, one north, the other south of the Alps, are compared in Table 45.

TABLE 45. ANOMALIES OF THE DURATION OF THE VEGETATIVE PERIOD AT TWO PLACES IN SWITZERLAND Z NORTH OF THE ALPS; M SOUTH OF THE ALPS

	Height m	Duration in days obs	standard	Anomaly obs — standard days
Z	355	221	243	−22
M	355	269	243	+26

The altitudes are—by chance—identical, so the two places have the same standard duration of 243 days. The actual duration has a deficit of minus 22 days north of the Alps; and a surplus of + 26 days south of the Alps. The numbers (Table 45) are really illustrative of the climate in these two zones. Everybody who crosses the Alps, especially in the late winter, is surprised by the climatic difference between north and south. He leaves the northern region with its ground everywhere covered with snow; but when he has passed through the Gotthard tunnel for instance, he arrives at the famous Lake of Lugano, where the trees are in full blossom.

The anomalies mentioned above give an exact picture of a sharp climatic divide between the places Z, north of the crest, and M, south of the crest at 335 m (1160 feet).

A place, Ri, 400 feet higher than M, has +25 days; and Br, 1400 ft higher, has an anomaly of +30 days. The three places have similar exposure towards the south, and are well protected

against invasions of polar air. The differences are surprisingly
small, in spite of the rather great differences in altitude. The
examples show that apparently the influence of altitude upon the
anomalies is practically eliminated. So it is possible to draw isanomals even in a region of great "relief-energy." The anomalies
combined with a standard element-height curve enable one to calculate the actual values for places where observations are available;
and, secondly, to interpolate with relatively small errors anomalies
and actual values for any point. It is beyond the compass of this
book to discuss the interesting picture resulting from application of
the method of anomalies and isonomals. One example is shown in
Figure 30, however: the isanomals of the duration of the vegetative
period in Switzerland. In spite of the great differences in altitude,

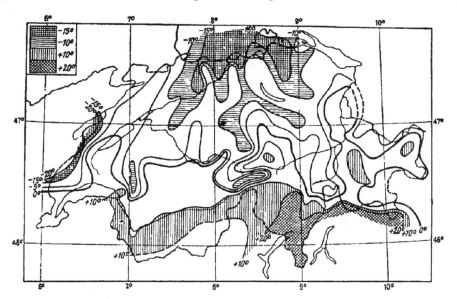

FIG. 30   Isanomals of the duration of the vegetative period in Switzerland
(After V. Conrad)

on the southern slopes, a large region of the Alps is characterized
by positive signs (a vegetative period longer than the standard
duration); and north of the Alps, a zone of negative signs is seen.
Not altitude but climatic exposure is represented.

There is no doubt that the method of isanomals is, up to the
present, the only method which eliminates the influence of altitude
to a high degree, and is, thus, the unique method of graphically
representing climatic conditions in a really mountainous region.

It goes without saying that this method is practicable for any variable climatic element.

XIV. 6. b. *Temperature and Geographical Latitude*

The best known *standard equation* which correlates the dependent variable, the temperature (*t*) with the independent vari-

TABLE 46. STANDARD DISTRIBUTION OF TEMPERATURE WITH GEOGRAPHICAL LATITUDE (After W Meinardus and Hann-Süring [5th Ed], Converted into °F)

ϕ	January	April	July	October	Year
N 90°	−42	−18	30	−11	−7 6
85	−36 6	−15 7	32 5	− 8 0	−6 2
80	−26 0	− 8 9	35 6	− 2 4	1.0
75	−20 2	− 2 2	38.1	6 8	5 5
70	−15 3	6 8	45.1	15 3	12.7
65	− 9 4	18 9	54 3	24 6	21 6
60	3 0	27 0	57 4	32 5	30.0
55	12 4	35 2	60 3	37.2	36 1
50	19.2	41 4	64 6	44 4	42 4
45	28 9	50 7	69 6	52 7	49.6
40	41 0	55 6	75 2	60 3	57 4
35	49 3	62 6	78.4	66 0	63 0
30	58 1	68 2	81.1	71 2	68.5
25	65 7	73 8	82 2	76 3	74 5
20	71.2	77 4	82 4	79 5	77.5
15	75 2	80 1	82 2	80 6	79 3
10	78 4	81 0	80 4	80 4	80 1
5	79.3	80.2	79 2	79 3	79.5
Eq.	79.5	79.9	78 1	79.7	79 2
S 5	79 5	79 7	76 8	78 8	78 4
10	79 3	78 6	75 0	78 3	77.5
15	78 6	77.4	72 1	75.9	75 9
20	77 7	75 2	68 0	73 0	73 2
25	75 7	71.2	63 5	69 1	69 6
30	71 4	65 7	58 5	64 4	65 1
35	65.7	59 4	53 2	59 5	59 0
40	60 1	54 5	48 2	53.1	53 4
45	54 1	46 4	43 2	46 4	47.8
50	46 6	43 3	37 9	41 7	42 4
55	41.0	35 2	27 7	33 4	34 3
60	35 8	27 5	15 6	24 8	25 9
65	30 7	19.0	3 0	16 2	17 2
70	25 7	7 5	− 9 4	6 1	7 5
75	19 8	− 6 7	−23 4	− 7.1	− 4.4
80	12 6	−19 8	−39.1	−22 0	−16 6
85	8.6	−28.7	−49.9	−28 5	−24 5
90	7.7	−33	−54	−31	−27.6

able, the geographical latitude (ϕ) is that of J. D. Forbes:[4]
$t = -17.8 + 44.9 \cos^2 (\phi - 6°30')°C.$

This equation is derived from observations only of the northern hemisphere, but is valid, despite this fact, approximately from 40°S to 60°N.

Such an analytical equation is needed for theoretical investigations. For practical purposes, such as calculating anomalies, Table 46 is preferable to an equation.

The values in the table[5] are probably the best available. Particularly, the formerly dubious temperatures of the high southern latitudes are replaced by those of W. Meinardus, who derived them from the observations of the most recent expeditions. The series of temperatures of January, July, and the year in the table were revised by Meinardus himself in 1936. The centigrade temperatures of the original table appear in this book converted into Fahrenheit degrees for the first time.

Naturally, the series of Table 46 should be interpolated for each degree, if used for calculating the anomalies of a great number of places.

In view of the great importance of the method of anomalies, Figure 31 has been inserted, in order to show how much a representation like this contributes to the understanding of the effect of ocean currents and continentality upon the temperature at the earth's surface.

XIV. 6. c. *Pleions, Meions*

Another application of anomalies and isanomals has been given by H. Arctowski in a series of papers.[6]

For a portion of the earth's surface, annual averages of an element are known for a normal period Anomalies are computed for each year of the period in question and represented by maps of isanomals. Regions encircled by a high positive isanomal are called *pleions* those characterized by high negative anomalies are named *antipleions* or *meions*. The method is applicable to every variable element. If temperature is dealt with, the terms *thermo-*

[4] James D. Forbes, "Inquiries about Terrestrial Temperature," *Transactions of the Royal Society of Edinburgh*, XXII (1859), pt. I, p. 75 ff.

[5] See Hann-Süring, *Lehrbuch der Meteorologie*, 5th ed., and V. Conrad, "Die klimatologischen Elemente ..," Köppen-Geiger, *Handbuch der Klimatologie*, IB, 123

[6] E.g., *L'Enchaînement des variations climatiques* (Bruxelles: Société Belge d'Astronomie, 1909); "Zur Dynamik der Klimaanderungen," *Met Zeit* 1914, pp. 417–426; *Communications de l'Institut de Géophysique et de Météorologie de l'Université de Lwów*. See also the valuable and critical review by Alfred Wegener, in Gerland's *Beiträge zur Geophysik*, vol. X, "Besprechungen," pp 298–299.

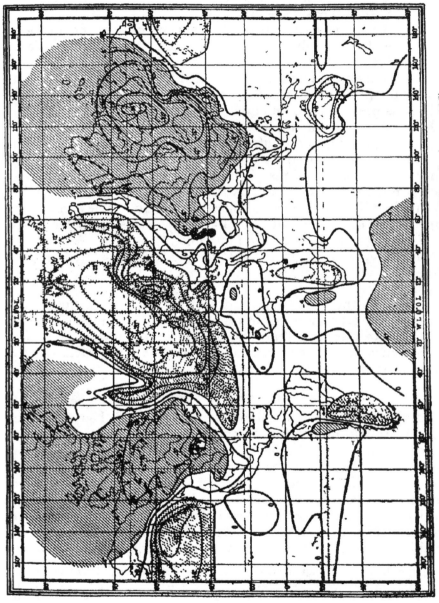

FIG. 31   Isanomals of temperature, January.   (After Hann-Suring, p. 185)

pleion and *thermomeion* are also used.   An interesting attempt has
been made to study the shifting of the pleions and meions from
year to year.   These efforts might be successful for the purposes
of dynamic climatology.   (XVIII. 6.)

XIV.   6. d. *Snow-Cover and Altitude*

The duration of snow-cover has been represented by the method
of isanomals for the eastern Alps and illustrates the influence
of a high mountain range extending east and west, upon snow
conditions.[7]

XIV.   6. e. *Duration of Sunshine vs. Elevation*

A study was made of the distribution of bright sunshine duration in the eastern Alps, by the method of isanomals.[8]   The map
(Fig. 32) shows that the positive anomalies are crowded on the
southern slopes of the Alps, and the negative on the northern.

Thus, the method of isanomals again gives the trend of the
effective climatic divide between the northern and southern slopes
of the Alps.

This result is of great consequence for the appreciation of the
method of isanomals, and runs parallel with the method of correlation.

XIV.   6. f. *Anomalies of Precipitation*

The knowledge of the geographical distribution of rainfall is
of the greatest importance for agriculture, hydrodynamics, architecture, etc.   Conversely, it would be difficult to find a branch of
human activity which is not interested—directly or indirectly—in
the amounts of precipitation which one may expect everywhere on
the earth's surface.   So it is no wonder that innumerable maps of
the distribution of precipitation appear in books, encyclopaedias,
journals, papers, and pamphlets.   The difficulties in drawing such
maps have not been overcome.

As long as the region for which the precipitation is to be shown
is flat, like a table, the use of *isohyets*, i.e., lines of equal precipitation, is more or less satisfactory.   If the country is hilly or even
mountainous, then difficulties set in, as in the case of temperature.

---

[7] V. Conrad, M. Winkler, "Beiträge zur Kenntnis der Schneedeckenverhältnisse
in den oesterreichischen Alpenländern," Gerland's *Beiträge zur Geophysik*, XXXIV
(1931), 473.

[8] V. Conrad, *Anomalien und Isanomalen der Sonnenscheindauer in den oesterreichischen Alpen* (Wien, 1938).

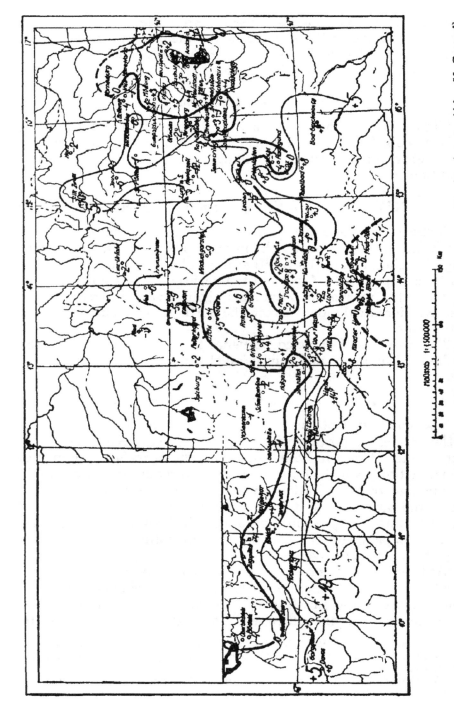

FIG. 32.   Isanomals of relative duration (% of possible) of bright sunshine in the eastern Alps in winter.   (After V. Conrad)

The comparison of like with like is not valid.   The difficulties are even much greater than with temperature.   With this last, there is great trouble with inversions in winter in the mountain valleys. Regardless of this real obstacle, the normal lapse rate (of temperature) does not vary much over the surface of the earth, if great regions are considered and details excluded.

For example: The lapse rate in Peru, 16°S, is 0.61 C°/100 m; in Ceylon, 7°N, 0.64; on the slopes of Mount Etna (Sicily), 38°N, 0.64; on Ben Nevis (Scotland), 57°N, 0.67 C°/100 m.

For precipitation, the conditions are of an entirely different nature.   For example: for Christmas Island (10°S, Indian Ocean, 190 miles south of Java) the author calculated the increase of rainfall to be 140 mm/100 m.   On the slopes of the volcano Tengger (Java), C. Braak found 97 mm/100 m, and that a *decrease*, between 600 m and 1700 m altitude.   On the slopes of Mount Idgen (Java), an *increase* of 260 mm/100 m was stated.

In a small valley a few miles long in the European Alps, the increase of precipitation, even with differences in height up to more than 7000 feet, varies from 14 mm/100 m to 163 mm/100 m (0.17 in./100 feet to 1.94 in./100 feet).   In every mountain range of temperate latitudes a decrease of precipitation with increasing altitude occurs.

Another striking example of these conditions is presented by O. Lütschg.[9]   In the small region of the famous Mont Cervin (Matterhorn), the following variations of precipitation with height are observed:

Pair of Stations		Variation of Precipitation with Height mm/100 m
Visp—Zermatt .	increase	9 3
Grachen—Zermatt	increase	912 0
Visp—Grachen	decrease	5 8
Visp—Saas Fee	increase	20 8
Saas Tamatten—Saas Fee .	invariance	0

In the Rhone valley (Wallis), the conditions mentioned are even more exaggerated:

[9] Otto Lütschg, "Ueber Niederschlag und Abfluss im Hochgebirge," *Veröffentlichung. der hydrologischen Abteilung der Schweizerischen Meteorologischen Zentralanstalt in Zürich* (Zürich, 1926), p. 137).

Pair of Stations		Variation of Precipitation with Height mm/100 m
Martigny-Ville—Sion	increase	347
Sion—Sierre .	decrease	200
Sierre—Visp	increase	84
Reckingen—Oberwald	increase	2252
Oberwald—Gletsch	increase	46

In view of these conditions no reduction to a fixed level is possible. Therefore, the usual rain map is the representation of *actual isohyets*.

What was said about actual isotherms is true also for actual isohyets, owing especially to the great irregularity of the phenomenon in regard to time and locality. The difficulties become very complex, but mostly in hilly and mountainous regions. Here again, the method of the isanomals is the only one that is able to eliminate the influence of height and to represent the true features of rainfall distribution. This method determines, in this case, too, the position of the climatic divide, even if it is shifted toward the lee side. The reason for this phenomenon belongs to the realm of physical climatology.

XIV. 7. PRECIPITATION PROFILES

In view of the great difficulties with isohyets, profiles of precipitation drawn through the mountain ranges are, at least theoretically, a good expedient;—theoretically, since the profiles can be drawn successfully only if the observations of a sufficient number of suitably located stations are available.

The principal requirement is that there be stations at or close to the points of the profile extremes. If this condition is not satisfied, the drawing of the profile is nothing but a guess, and the profile itself represents fancy more than truth. The stations should not be too scattered on either side of the location of the profile.

The network of stations up to the crests and summits of the high mountains is very dense in Switzerland, so an example from the western Alps may be useful. This is reproduced in Figure 33, where Lütschg[10] gives a profile through the Bernese Alps from NW to SE.

[10] O. Lütschg, *Ueber Niederschlag und Abfluss im Hochgebirge* (Zürich, 1926), p. 138.

This profile may serve in some points as a pattern. The vertical scale is not so exaggerated as is often the case in such representations. It is 10.5 times greater than the horizontal. The length of the profile is 278 miles, with 27 stations. On the average,

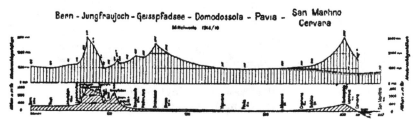

FIG 33.   Precipitation profile across the Bernese Alps from NW to SE.
(After O. Lutschg)

there is one station every ten miles. This ratio does not give the right picture, because, deliberately, the stations are not evenly distributed over the profile. The stations are very dense within the region with greatest differences in height. There, 10 stations are located within 37 miles. In the plain of the Po (Italy), two stations are shown within 60 miles.

It is usual to blacken the mountain profile. In the figure, it is slightly hatched so that certain levels can be indicated by horizontal lines. This gives valuable information.

The mountain and the precipitation profiles are drawn separately: the two scales do not overlap, but are arranged one above the other, with homologous points on the same vertical.

At the southern side (right), the profile is bifurcated, showing the amount of rain for the plain on the one hand, and for the hills (about 3000 ft) on the other.

The clear marks and lines showing the vertical and the horizontal scale, should not be overlooked.

This representation is well-planned, indeed. Yet one unfamiliar with Swiss orography, looking at the graph, might not understand why the amounts increase so much between the 100 km (see figure 33) and the 150 km point. Here, the profile is influenced by the contour features on either side.

Nevertheless, profiles based on sufficient and exact data are an excellent and necessary supplement to the maps of actual isohyets, never satisfactory in mountainous countries.

Figure 34 exhibits a profile of snowfall across New England from the New Hampshire coast to Williamstown, Mass. C. F.

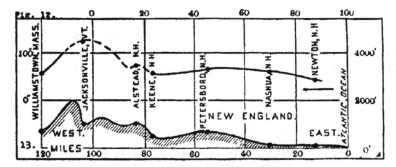

FIG. 34.   Profile of snowfall across New England from the New Hampshire
coast to Williamstown, Mass.   (After C. F. Brooks)

Brooks's graph [11] is, on the whole, in good agreement with the foregoing suggestions.   Representation of the hypothetical portion of the curve by a broken line might be generally adopted. An arrow indicates the direction of the snow bearing winds.

XIV.   8. VARIABILITY OF PRECIPITATION AND AMOUNTS OF PRECIPITATION

The independent variant is the amount of precipitation, the dependent is the variability.   This combination shows how adaptable the method of anomalies and isanomalies is.   The problem was discussed in III 8

[11] Charles F Brooks, "The Snowfall of the Eastern United States," *M W.R.*, January 1915, vol 43, p. 9

CHAPTER XV

WIND

XV. 1. Wind Roses

THE WIND ROSE is defined as "a diagram showing for a definite locality or district, and usually for a more or less extended period, the proportion of winds blowing from each of the leading points of the compass." [1] In the simplest form, eight directions are used, and relative frequencies (per cent) should be preferred. The frequency of calms is indicated by a circle in the center of the rose. The radius is proportional to the frequency; or, as on United States pilot charts, a number in the circle shows the per cent of calms. [2] The method is too well known to require further explanation.

Sir Napier Shaw constructed a composite wind rose or wind "star" in the utmost detail. [3] This is reproduced in Figure 35. It shows the annual course of the frequency of three classes of wind velocity for 16 directions.

The three steps of velocity are:

1) light winds
2) moderate or strong winds
3) gales

Light winds (Beaufort 1 to 3), are represented in the rose by a single line. The second group includes the Beaufort degrees 4 to 7, and is indicated by a double open line. At their ends are small blackened sections which represent the number of gales of force 8 and more. Frequencies are given in per cent. "The indentation of the base line allows us to distinguish the information for the summer half, April to September, for the northern hemisphere, from the two winter quarters, January to March on the left for a spectator looking from the centre and October to December on the right. Small letters *A* and *S*, hardly visible in the reproduction without a glass, mark the ends of the salient." Thus, one "star"

[1] *Meteorological Glossary* (3d ed , London, 1939)
[2] For a good example of wind roses and their applications see Glenn T. Trewartha, *An Introduction to Weather and Climate* (2d ed , New York, 1943), p 112
[3] Sir Napier Shaw, *The Drama of Weather* (Cambridge, 1933), p. 165.

replaces twelve normal wind roses. Drawing one star may be more troublesome than drawing twelve simple roses, but much printing space is saved. Wind roses are often used to indicate the geographical distribution of wind conditions, and are legible even on a rather small-scale map; but this is not true for the "stars."

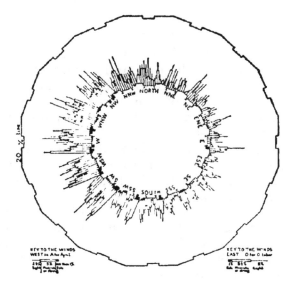

FIG. 35 The wind "star" (composite wind rose)
(After Sir Napier Shaw)

For climatological purposes, the data used for wind roses are better given in numerical tables, which take into consideration also the frequency of calms. Here, too, percentages are preferable to absolute frequencies.[4] The horizontal entry of these tables of distribution of wind directions has headings for the eight directions and the calm; the vertical is used for the months. Such statistics are indispensable.

XV. 1. a. *Wind Roses for Different Elements*

Wind roses for different elements are based on the principle mentioned, and are illustrated by the example in Table 47.[5] The

[4] See models in R. DeC. Ward and C. F. Brooks, "Climates of North America," Köppen-Geiger, *Handbuch der Klimatologie*, pt J, pp. 234 ff.

[5] After A. Wijkander, "Observations météorologiques de l'expédition arctique Suédoise 1872–73," *Kongl. Svenska Vetenskaps-Akad Handlinger Bandet 12*, no. 7 (Stockholm, 1875).

TABLE 47   WIND ROSES FOR DIFFERENT ELEMENTS OBSERVED AT MOSSEL BAY, SPITSBERGEN (80°N, 15°E), 1872/73. Deviations for Temperature ($\Delta t°C$) and Cloudiness ($\Delta C\%$), and Per Cent of Probability of Precipitation (P-Pro %) (After A Wijkander)

Wind direction	$\Delta t°C$	$\Delta C\%$	P-Pro %
N . .	−1 8	+ 9	33
NE . . .	−2 4	− 4	18
E .	−2 3	−17	8
SE	+0.5	−13	5
S. ...	+4.9	− 6	6
SW	+3.8	+ 3	19
W.. .	−0 4	+14	40
NW. . . .	−2 0	+16	38
Calms...... .	−1 1	−10	11

NOTE. To avoid misunderstanding The "P-Pro" percentage relates to single wind directions, e g , there are 33 rainy days out of 100 days with N-wind, and only 5 rainy days out of 100 with SE-wind The representation clearly shows that a strong tendency toward precipitation exists with northerly and westerly winds, and a small one with southerly and easterly winds

numbers in the columns $\Delta t°C$ and $\Delta C\%$ denote average deviations for the eight directions and for calm. "P-Pro" means the average probabilities of precipitation. The method combines the immediately observed surface winds with the synchronous values of the different elements.

Wind roses are frequently a good climatic characterization, but do not contribute much to air-mass analysis. The correlation between the momentary surface wind and the life history of an air mass is not well defined.

XV. 2. MAPS OF STREAMLINES

For a cartographical representation, maps of *streamlines* seem to offer the best climatological information.

Streamlines exhibit the state of air motion at a particular moment, and must be carefully distinguished from trajectories or actual paths of an air particle Streamlines reproduce the synoptic picture at a given moment, trajectories the chronological story of an air particle. Maps of streamlines should be accompanied by maps of the synchronous distribution of air pressure.

The technique of drawing streamlines includes the consideration that the resultant wind direction at a place should be the tangent to the streamline. Besides, physical considerations concerning the field of pressure and temperature are of primary interest in drawing streamlines.

As was stated in Chapter VIII, the observations and publications include either (a) direction and velocity, or (b) only a frequency distribution. In the first case, the resultant wind velocity and steadiness (an important characteristic of wind conditions) can be computed for every station.

Climatological streamlines represent, of course, average conditions. As far as the wind is concerned, a climatological streamline does not mean a reality like a meteorological streamline: the distinction is the same as between an arithmetical mean and an actual value. The arithmetical mean becomes the more characteristic the smaller the variability. At a tropical place with a very small temperature variability the monthly mean of temperature is no longer a fictitious number. In wind conditions "steadiness" replaces "variability." The higher the steadiness, the more the average streamline approximates actuality. Therefore, it is of great interest to add the representation of steadiness to that of velocity and direction on a climatological map of streamlines

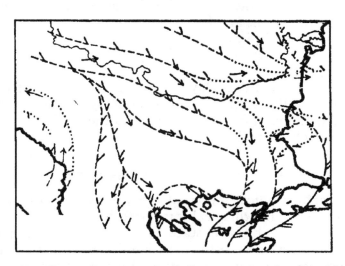

FIG. 36   Streamlines over the Balkans, average for July 1917.
(After E. Kuhlbrodt)

As an example, average streamlines are given for the eastern Balkans, according to observations made in July 1917. The special method applied to the construction of this map (Fig. 36), has been elaborated by Alfred Wegener, and fulfills apparently all requirements. The trend of the streamlines in Figure 36 is interesting, in that the graph clearly shows how the general west-east streaming of the air over the higher latitudes of Europe is deflected to winds

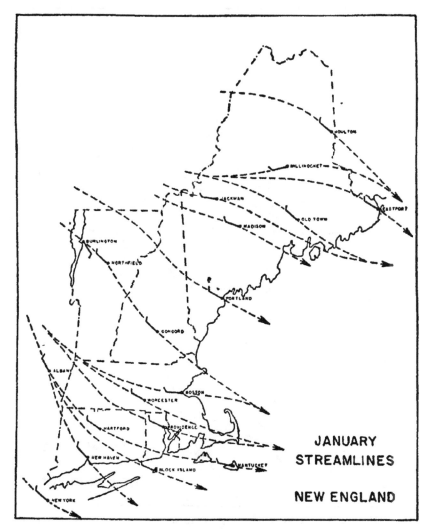

Fig. 37a. Streamlines of New England in January. (After V. Conrad)

with a northern component; this is more and more strengthened towards south and east. So the map constitutes a representation of the "Etesians." The feathers, marked directly on the stream-lines, indicate the velocity of the wind in relative measure. In the present map, half feathers are equivalent to the weakest winds, and three full feathers indicate a very strong average velocity. (See the region off the coast of Asia Minor.)

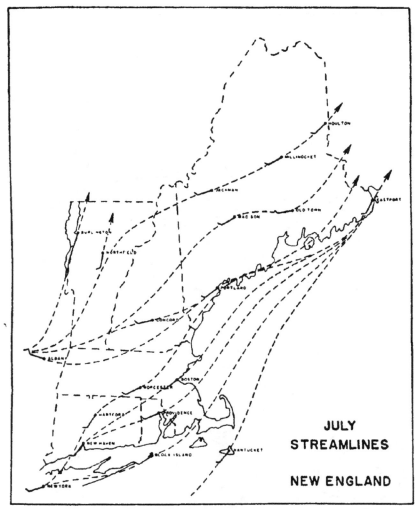

FIG. 37b. Streamlines of New England in July. (After V. Conrad)

The steadiness is expressed in the following way:

1) If the streamlines are dotted, the steadiness is between 0% and 33%
2) Broken streamlines mean a steadiness of 33% to 66%
3) Full streamlines indicate a steadiness greater than 66%

From the present representation, one can see that the northerly winds of the Aegean Sea are not only the strongest winds but also those with greatest steadiness.

XV.   2. a. *Streamlines Based Only upon Frequencies of Direction*

If records of velocities are lacking, average resultant wind directions may nevertheless be computed, as was proved earlier (see VIII. 2). Then, of course, only streamlines can be constructed; naturally no velocities, no steadiness can be indicated. Even a plain streamline picture is climatically instructive, as is shown in Figures 37 a and b. At first glance, the reader sees the relatively slight shifting, even if climatically decisive, of the average wind from winter to summer. The continental features of winter on the New England coast clearly appear, and the warm, humid air masses which overflow the country in summer are indicated.

CHAPTER XVI

AIR MASS CLIMATOLOGY

AVERAGE STREAMLINES (mentioned in the foregoing chapter) are closely connected with the concept of air masses, so that this branch of climatology, rich in prospect, may be discussed here. The methods used in these investigations are similar to the usual statistical procedures. They start from the frequencies of different types of air masses for certain parts of the year. An air-mass calendar taken from the daily weather maps is the foundation Relative frequencies should be preferred.

One of the results is also of special methodological interest: Air masses can be divided into two main groups: continental (c), and maritime (m). The frequency-ratio c/m is obviously a good characteristic of the climate, and perhaps a better index of continentality than the factors of continentality (discussed later on).

Table 48 offers two examples of the seasonal ratios c/m, one for western Europe,[1] the other for the eastern United States.[2]

TABLE 48. DEGREE OF CONTINENTALITY DERIVED FROM FREQUENCIES OF CONTINENTAL (c) AND MARITIME (m) AIR MASSES (FOR PENNSYLVANIA AND FOR WESTERN EUROPE)

	Winter	Spring	Summer	Fall
	PENNSYLVANIA			
c/m	1.65	1 08	1 02	1 07
	WESTERN EUROPE			
c/m	0 58	0 40	0 18	0 25

The results of Table 48 do not need further explanation. In the eastern United States, especially in winter, the continental

[1] E Dinies, *Luftkorper Klimatologie* (Frankfurt a M , 1932) (air mass system proposed by F. Linke).

[2] H Landsberg, "Air Mass Climatology for Central Pennsylvania," Gerland's *Beiträge zur Geophysik*, vol. 51 (1937) p 263 (H C. Willett's classification). For further information see Jerome Namias, *Air Mass and Isentropic Analysis* (5th ed , with contributions by T Bergeron, B. Haurwitz, G. Miller, A K Showalter, R G Stone, and H C. Willett). In this connection, chapter II on "Classification of Air Masses," with list on p. 76, is of particular interest See, for an informative survey, chapter IV, "Air Masses and Fronts," in Glenn T Trewartha, *An Introduction to Weather and Climate* (2d ed , New York, 1943), pp 190 ff

component predominates; in western Europe, the maritime.   The first result is confirmed by the average streamlines in winter over New England (Fig. 37).   The contrast between the two climates is well described by the ratio c/m.

The great merit of the air mass index of continentality is that it needs no special assumption or hypothesis.   Thus, it yields incontestable results.   The disadvantage is that the basic air mass calendar entails much work and critical study.

CHAPTER XVII

NUMERICAL CHARACTERIZATION OF DIFFERENT CLIMATIC FEATURES

XVII. 1. CONTINENTALITY

THE INCREASE of the annual range of temperature inland is perhaps the most striking effect of the continental surface on climate. Conversely, the annual range (A) is taken as a measure of the climatic factor, called *continentality*. Because a general physical correlation exists between annual range and geographical latitude, the range has to be reduced to equality for all latitudes; thus the expression

$$\frac{A}{\sin \phi}$$

is the measure sought for.

The formula, now frequently used, reads:[1]

$$C = \frac{1.7 \times A}{\sin \phi} - 20.4$$

where C is the coefficient of continentality in per cent, A the annual range of temperature in centigrade degrees, and ϕ the geographical latitude.

Example:

    1) Jakutsk (62.0°N, 129.7°E, 320 ft) $A = 61.6$ C°
    2) Thorshavn (62.0°N, 6.8°E, 83 ft.) $A = 7.6$ C°

Therefore

$$C_{\text{Jak}} = \frac{1.7 \times 61.6}{0.88} - 20.4 = 98\%$$

$$C_{\text{Thors}} = \frac{1.7 \times 7.6}{0.88} - 20.4 = -6\%$$

The second example shows that the constants have to be somewhat changed, since a "negative continentality" has no physical meaning. If the continentality is zero, the climate is no longer influenced by continental surfaces in any way. Consequently, it is always better to speak of continentality than of "oceanity."

[1] Wl Gorczynski, "Sur le calcul du degré du continentalisme et son application dans la climatologie," *Geografisker Annaler*, 1920, p 324.

From the methodological point of view, the fact of a negative coefficient is not so important and easily corrected; the later comments by Gorczynski himself and by O. V. Johansson are valuable in this respect. The really weak point, however, has (to the best of the writer's knowledge) never been emphasized, simple as it may be.

The coefficient of continentality C becomes infinite ($C = \infty$), or practically infinite, in the interior tropical belt, if sin $\phi \doteq 0$. Therefore, all these formulae, going back to W. Zenker's investigations, are no longer valid for very low latitudes. This point is fundamental in regard to method, and shows that one must always take into account the boundary conditions of the formulae used.

On the other hand, it should not be forgotten that the formula gives a comparably numerical representation of the continentality, although for only a part of the earth's surface.

Discussing a formula of continentality given by R. Spitaler and mapped by G. Swoboda, D. Brunt [2] arrives at an equation which shows a clearer physical significance than that mentioned. His formula reads:

$$n_e + 0.12 = \Delta t / 130.61 \cdot \Delta S$$

where n_e denotes the *continentality-factor* of a given place, Δt the annual range of temperature, and ΔS the annual range of the average intensity of solar radiation in the latitude of the place in question.

Then, "($n_e + 0.12$) measures the response in monthly mean temperature to a unit change in the mean intensity of solar radiation."

We have only to add that D. Brunt alludes to the point that in the maps of Swoboda no isolines are drawn within the zone 20°N to 20°S, "on account of the uncertainty in the computation of n_e for this region." This remark is in good agreement with the criticism above mentioned.

XVII. 2. LIMITS BETWEEN TIMBER-FOREST, STEPPE, AND DESERT. EFFECTIVENESS OF PRECIPITATION

Supan and Köppen derive numerical limits of their climatic provinces from average data of temperature and precipitation. For example:

1) The coldest month in the "tropical (A)-Climate" is warmer than 64°F.

[2] David Brunt, "Climatic Continentality and Oceanity," *Geographical Journal*, London, vol. LXIV (1924), p 43.

2) The highest average monthly temperature within the "regions of snow and frost" (EF-Climates) is below 32°F., etc.

Supan had already defined the tropical zone by the annual isotherm of 68°F. This definition marked a great advance in this subject. That was also Köppen's opinion, although his numerical limits are incomparably more practical.

Great difficulties arise if the effect of precipitation has to be estimated. Chief of these is the problem of the limit between steppe and timber-forest on the one hand and steppe and desert on the other.

The consequences of a given amount of rain are dependent on temperature, owing to the effect of evaporation. It is impossible, either theoretically or practically, to formulate an equation which would take account of all effective elements and factors, such as radiation, wind, soil, continentality, humidity, elevation, etc. Therefore, Köppen had to be satisfied with rough empirical approximations.[3] He himself frequently changed these formulas, so that one example may suffice. Often a slight modification of Köppen's rules helps if a special region has to be described.

The effectiveness of precipitation is greater at low temperatures. Consequently, for instance, one must discriminate between steppes with precipitation in cool winter and steppes with precipitation in hot summer.

The limit between a forest climate and a steppe climate is indicated by the following relations between temperature, t (°C), and precipitation, r (cm):

a) rainy period definitely in winter: $r = 2t$
b) rainy period definitely in summer: $r = 2t + 28$
c) no definite rainy period: $r = 2t + 14$

The limit between steppe and desert can be estimated to lie where the rainfall is one-half of the amount stipulated above, in each case:

a) winter rain $r = t$
b) summer rain $r = t + 14$
c) transition cases $r = t + 7$

The following example may be instructive:

[3] For further details, see W Köppen, *Grundriss der Klimatologie* (2d ed., Leipzig, 1931), and Koppen-Geiger, *Handbuch der Klimatologie*, vol. Ic (1936).

1) Stalingrad (48.7°N, 44.5°E,   138 ft): $t = 7.7°C$; $r = 37$ cm
2) Achtuba   (48.3°N, 46.9°E,   11 ft): $t = 7.7°C$; $r = 25$ cm
3) Astrachan (46.4°N, 48.0°E, −30 ft): $t = 9.4°C$; $r = 15$ cm

Astrachan lies in the delta region of the Volga river, Achtuba is 150 miles distant in the Volga basin, and Stalingrad lies another 120 miles up the river.   For the present purpose one can assume in rough approximation that the distribution of precipitation over the year is more or less uniform.   Therefore the formulas (c) should be valid.   For the temperature of Stalingrad and Achtuba we get the formula:

$$\text{steppe climate if } r < (2t + 14) = 29 \text{ cm}$$

Therefore Stalingrad with 37 cm has a forest climate: Achtuba with 25 cm has a steppe climate, and the climate of Astrachan is that of the desert, because there is less than 16 cm rainfall.

The boundary between forest and steppe climate runs between Stalingrad and Achtuba.   The border line of the desert can be assumed to lie between Astrachan and Achtuba, but closest to Astrachan.   An interpolation regarding the location of such a border line between two places should be made with the greatest caution.   The reader may be reminded of coherent and non-coherent climatic regions.   Every interpolation presumes continuity of the variations.[4]

The index of precipitation effectiveness given by C. W. Thornthwaite [5] reads:

$$I = 115 \left(\frac{P}{T - 10} \right)_n^{10/9}$$

The symbol I is called the *precipitation-effectiveness index*. P means the monthly precipitation in inches and T the average monthly temperature in F°, while n indicates the number of the month in question.

There is no doubt as to how to calculate this index, but it is a rather wearisome task.   The reader who intends to use this formula finds some help from the nomogram given by Thornthwaite.

[4] Since the stations are generally far distant from one another, the interpolation of the values, critical for the location of the border lines between forest and steppe and between steppe and desert, is not an entirely reliable procedure.   Therefore, botanical facts also ought to be considered in drawing these border lines

[5] C. W. Thornthwaite, "The Climates of North America According to a New Classification," *Geographical Review*, October 1931,—"The Climate of the Earth," *ibid.*, July 1933;—*Atlas of Climatic Types in the United States 1900–1939* (Washington, 1941).

PART IV

THE CLIMATOGRAPHY

CHAPTER XVIII

ARRANGEMENT OF A CLIMATOGRAPHY

IN THE EARLIER PARTS of this book, methods of examining observational data in a critical and quantitative way were explained. From the homogeneous and reduced series of elements, characteristics were derived by means of simple mathematical statistical methods.

The idea which runs right through is to proceed from the qualitative to the quantitative, and to arrive at comparable and reproducible results. These are the methods of analysis.

But a climatography should give more than this. It should offer a full picture of the climate in question. This aim can rarely be reached. Nevertheless, the content, outlines, and arrangement of a more or less complete climatography are here suggested.

XVIII. 1. THE INTRODUCTION

The Introduction should contain a short geographical description of the region in question, illustrated by a clear, legible, orographic map. At least the trends of rivers, streams, and lakes, as well as those of the mountain ranges, should be outlined.

A second map showing the locations of the stations should not be omitted The reader finds a pattern of the latter in each section of the Kóppen-Geiger *Handbuch der Klimatologie*, e g, in R. DeC Ward and C. F Brooks, "Climates of North America." But a net of parallels and meridians running right across the map should not be forgotten. An alphabetical list of stations is indispensable. The three geographical coordinates should be added to each station name, with a number referring to the map of stations.

The second part of the introduction should give a survey of the material in connection with the map of stations. The kind of shelters and rain gauges and their exposure at the majority of stations should be briefly and clearly described. The history at

least of the "normal stations" should be given, pointing to chang-
ings of exposure of the instruments, of methods of observation, of
operators, etc.

Smoothing of certain series correcting obviously false observations and interpolating of missing values should be indicated.

The series whose relative homogeneity has been examined
should be given *in extenso*, as well as detailed results of these computations with an exact indication of the methods applied. Simi-
larly, a clear report with all necessary numerical additions should
be given regarding the reduction to the chosen "normal period."
It does not suffice to make a general statement that the series are
homogeneous and that they are reduced to a given period. Only
from detailed numerical results can the reader judge how far the
representation is reliable and useful.

XVIII. 2. STATIC CLIMATOLOGY

The static climatology deals with the average state of the
atmosphere. It is more or less identical with climatological
statistics. Therefore, the backbone of this part is represented by
material which must be given in numerical tables and graphs.

There are different kinds of tables.

1) The *climatic table* contains average values for a certain
place, derived from a certain period, and besides, extreme values.
Samples of this kind of table are given in the appendix. The title
of each climatic table must show, also, the name of place and
country, and the three geographical coördinates.

It often happens that it is impossible to give all the data for
one and the same period. Then the period must be mentioned at
the head of its column, which, in any case, has to contain the units
of the respective element. (°C, °F, tercentesimal degrees or absolute degrees, "°A," or "°K" = Kelvin; inches or millimeters,
millibars, millimeters of mercury or inches of mercury, meters,
feet, etc.).

2) The climatic table can also be replaced by tables each of
which is related to a certain element (average temperature, cloudiness, etc.) and contains a list of stations and the averages of the
element in question for the months and for the year. As a sample,
again the tables in "The Climates of North America" (R. De-
Courcy Ward and C. F. Brooks) can be recommended. Many
authors prefer tables of this kind because they save space and
printing costs. Climatic tables should be given, in any case, for

the most representative places of the region.   If a climatography
of New Hampshire were composed, climatic tables would have to
be given for Mount Washington, for a suitable station at its base,
and for at least one station near the sea.

3) For stations with more than ten years' operation, tables
should be included which contain the monthly and annual means
or sums, at least, for temperature, rainfall, and pressure, for all
available months and years.   Cloudiness should be added, if in
any way possible.[1]

Whatever kind of climatic tables is considered, frequency distributions (see II. 4 and ff.) of the various elements contribute
essentially to the quantitative description of climate and can be
advantageous in many even not specifically climatological problems   Recently J. Namias ("Construction of 10000-Foot Pressure Charts Over Ocean Areas," BAMS vol. 25, 1944, p. 177 ff.)
has used frequency distributions of air temperature by 5° squares
over the Atlantic Ocean for his forecast technique.

XVIII. 3. Monographs

Monographs are of great value, inasmuch as they deal with
long homogeneous series of different elements obtained at a place
representative of a greater region.   It is impossible to give specific
instructions for composing such investigations and publications.
The more details and statistics are given, the more useful the
monograph is.   The reader may be referred to some models of
different kinds:

1) Under the direction of Dr. J. B. Macelwane, S.J., there
appeared *Meteorological Means and Normals from Observations at
Saint Louis University, 1911–1935*, by John J. Renk, S.J., two
volumes which offered only numerical tables, extremely useful for
other scholars.

2) A. Wagner, in *Der Jahrliche Gang der meteorologischen Elements in Wien* (Wien, 1930), presents a full discussion and a great
collection of statistical tables.   Both monographs are restricted to
the annual course of the elements.

3) F. Steinhauser, in *Die Meteorologie des Sonnblicks* (Wien,
1938), gives the pattern of a monograph which considers the annual

---

[1] For examples see R DeC Ward and C F Brooks, "Climates of North America,"
in Köppen-Geiger, *Handbuch der Klimatologie*, p 266 ff, or U. S Weather Bureau,
"Climatic Summary of the United States," e g , W M. Wilson, Section 82, South Central
New York, p. 3 ff.

course as well as the daily, in an exact way, by numerical data and their discussion.

A different kind of monograph is that of O. L. Fassig, *The Climate of Baltimore*. It will be mentioned later (XVIII. 7.)

XVIII. 4. Records of Self-registering Instruments

If continuous records are at hand, the daily course of the respective elements should be shown. Often, two-hourly values are sufficient. In the first place, the daily variation has to be reproduced numerically. Graphs are always welcome but do *not* replace numerical tables. From the tables the reader can make himself a correct graph. The converse is not true.

From the continuous records *periodic* and *aperiodic* variations have to be derived.

The periodic variation is the difference between the highest and the lowest value of the average daily course (perhaps for the period of a month) of the element (pressure, temperature, relative humidity, etc.).

The aperiodic variation is obtained from the average daily extremes. In the case of temperature, it can be obtained not only from the records of the thermograph but also from the daily readings of the extreme thermometers.

XVIII. 5. Change of Temperature and Precipitation with Height

In a mountainous country, the variation of temperature with height has to be carefully considered, and its annual course, in any case, must be presented. Here also numerical tables are preferable to graphs alone. The same is valid for the variation of precipitation with height.

If the number of stations is not too small, and there are considerable differences in height, a standard temperature-height curve and a precipitation-height curve should be constructed. From these curves, the anomalies of different places are calculated.

XVIII. 6. Graphical Representation

As far as graphical representation is concerned, it is a matter of taste which and how many Cartesian diagrams are shown. If these appear in addition to numerical tables, they are practical and useful.

The geographical distribution of at least the most important elements (pressure, temperature, precipitation, cloudiness) should be represented by maps of isolines, provided the country is not of too great "relief energy" and the mountains do not occupy too large a portion (see details XIV. 2).

For a really mountainous country with a dense network of stations up to great altitudes, attention should be focused on the construction of maps of *isanomals*. Only these maps are able to reveal the real distribution of temperature and precipitation, and particularly the climatic divides, in a mountainous country. The method of isanomals eliminates the influence of height on the element in question and permits a reasonable interpolation for places where no data are available. The anomalies, together with the standard curve for the region, yield the actual values of the element (see XIV. 2).

Maps representing the data of the beginning and the end of well-defined seasons contribute much to the knowledge of a climate. Examples are: season with frost, vegetative period, etc. (See VI. 2, 3.)

The addition of an exact definition of the season represented should never be overlooked. The *frost period*, for instance, can be defined in different ways:

1) as the period between the dates when the minimum temperature drops below the freezing point the first and the last time in the cold season.
2) as the period between the dates of the first and the last day with a *mean* temperature = 32°F.
3) as the period between the dates at which the average curve of the annual course passes the freezing point.

The *vegetative period*, also, can be defined variously, as follows:

1) as the interval of time which lies between the dates when the average annual course passes 43°F.
2) as the space of time between the last and the first "killing frost"; this period is called "growing season" in the United States.

The number of these examples could be greatly increased. Therefore a clear, exact definition is essential.

As we have already pointed out, the dates of the beginning and the end of a period yield the *duration of the period*. The cartographical representation of "durations" is valuable. Gener-

ally they are calculated in days. For use in maps they should
commonly be converted into weeks. These correspond better to
the usual degree of accuracy of the data and particularly to that

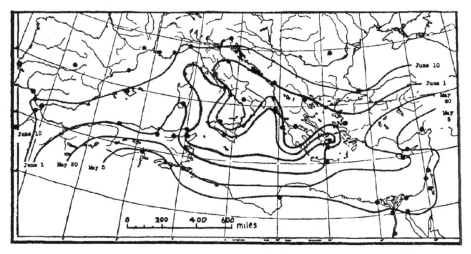

FIG. 38. Isolines of the dates at which the curve of the annual course
of temperature rises to 70°F in the Mediterranean region. (After V. Conrad)

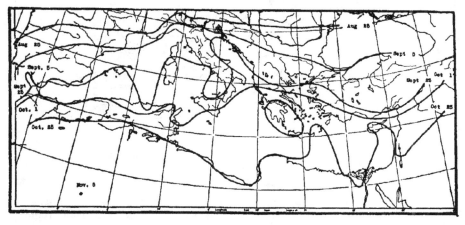

FIG. 39. Isolines of the dates at which the curve of the annual course of
temperature drops below 70°F in the Mediterranean region.   (After V Conrad)

of the trend of the respective isolines.   In some cases the unit of a
month yields the best and the most reliable isolines.

Figures 38 to 41 illustrate the method of representing the
dates of beginning and end of defined periods, and the duration

of such periods.   (See also XIV. 5. a.) [2]   In Figure 38, isolines of
the dates are seen, on which the curve of the annual course of
temperature reaches 70°F in the Mediterranean region.   These

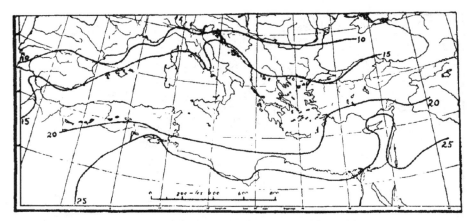

FIG 40   Isolines of the duration in weeks of an average daily temperature
$\geqq$ 70°F in the Mediterranean region   (Duration of the hot season)   (After
V Conrad)

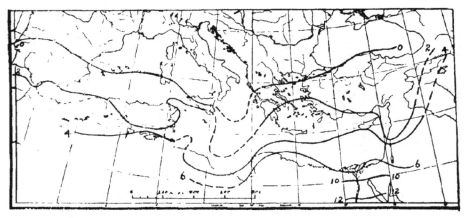

FIG. 41.   Isolines of the number of months with an amount of rain $\leqq$ 0.2
inch   (Isolines of the length of the dry season, Mediterranean region).
(After V Conrad)

isolines are characteristic of the beginning of the hot period; the
next, Figure 39, shows its end in the same way.   On the two maps,
the average migration of the chosen threshold of temperature can

[2] S. S Visher, "The Seasons' Arrivals and Lengths," *Annals of the Association of
American Geographers*, vol. 33, 1943, pp 129–134, gives similar maps for the United
States

be observed. The average temperature reaches 70°F in North Africa on May 5, migrates northward and arrives in southern Europe, June 10. The retarding effect of the water basin itself is well emphasized by means of the closed isoline which corresponds to June 5.[3]

Figures 40 and 41 give examples of durations. The first of them is derived from Figures 38 and 39, and represents the duration in *weeks* of temperatures equal to or higher than 70°F. It is the *duration* of the hot season in this region. On the average, the duration increases from north to south. The intensity of this increase is, however, greatly diminished by the various effects of the surface of the water.

The discussion of a map of well drawn isolines can be very stimulating and can result in a surprising insight into the regional climate.

The last example for this group of problems is offered by Figure 41, representing the length of the dry season in the Mediterranean region. Three features of this map may be stressed:

1) The "dry season" is well defined by the indication: Number of months with an amount of rain ≤ 0.2 in.

2) Parts of the isolines, the trends of which are only assumed, not really based on observations, are drawn as broken lines.

3) The month is taken as a unit, because of the high variability especially of small amounts of rain, and the great uncertainty in measuring such tiny quantities, particularly at times of high temperature and great evaporation.

XVIII. 7. DYNAMICAL CLIMATOLOGY

This expression was coined by Tor Bergeron, when he extended the meteorological methods of the Norwegian school to climatology.[4] Static or statistical climatology represents the average and extreme *states* of atmospheric conditions. Dynamical climatology means perhaps *development* of the average states. No doubt many features of each division overlap. Usually, every old-fashioned climatography contains, for instance, variabilities which belong to the dynamic division.

[3] This date which belongs to this isoline was unfortunately omitted in the figure
[4] See review by H. C. Willett, "Ground Plan of a Dynamic Meteorology," MWR Vol 59, (1931), 219–223.

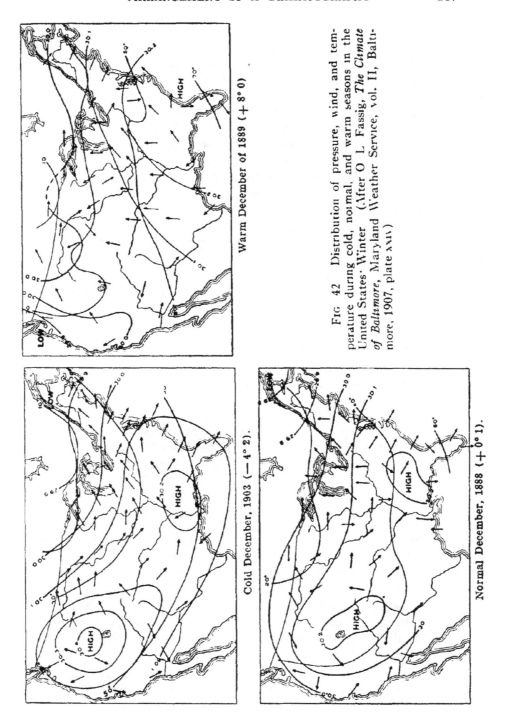

FIG. 42 Distribution of pressure, wind, and temperature during cold, normal, and warm seasons in the United States. Winter. (After O L Fassig, *The Climate of Baltimore*, Maryland Weather Service, vol. II, Baltimore, 1907, plate xxiv)

Warm December of 1889 (+ 8° 0)

Cold December, 1903 (− 4° 2).

Normal December, 1888 (+ 0° 1).

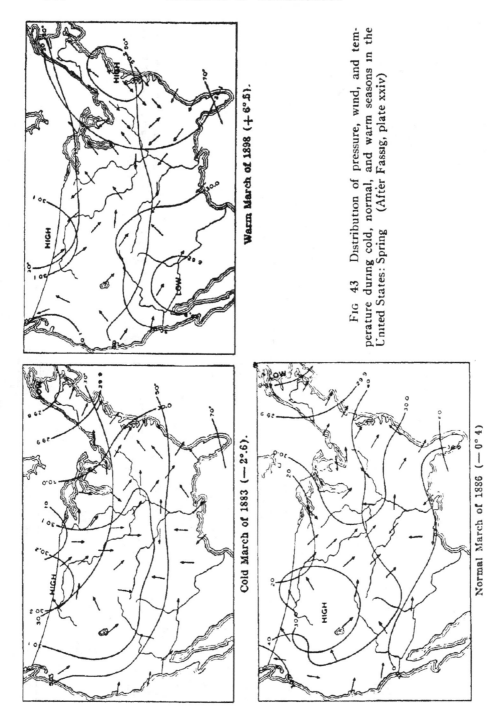

Warm March of 1898 (+ 6°.5).

Cold March of 1883 (— 2°.6).

Normal March of 1886 (— 0°.4)

FIG 43 Distribution of pressure, wind, and temperature during cold, normal, and warm seasons in the United States: Spring (After Fassig, plate xxiv)

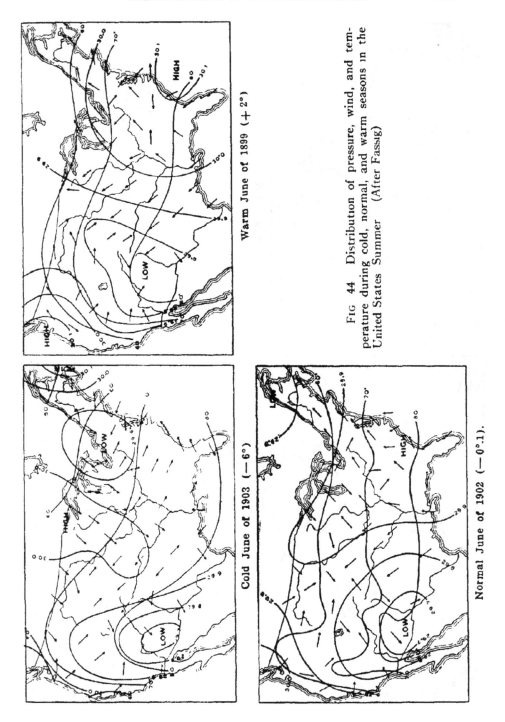

Warm June of 1899 (+ 2°)

Cold June of 1903 (− 6°)

Normal June of 1902 (−0°.1).

FIG 44 Distribution of pressure, wind, and temperature during cold, normal, and warm seasons in the United States Summer (After Fassig)

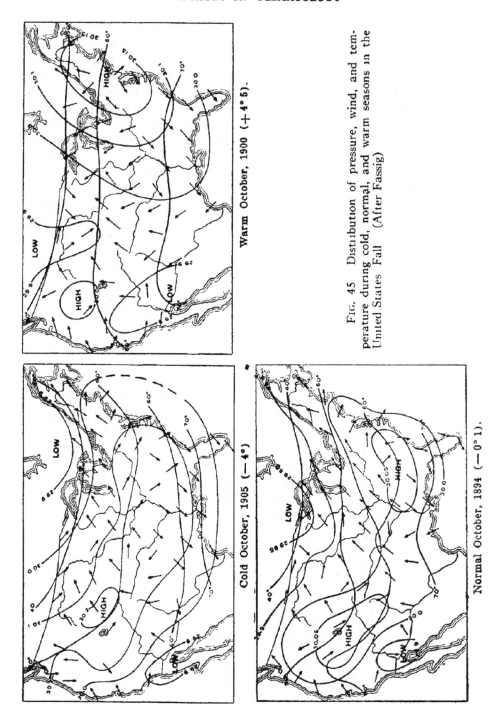

FIG. 45 Distribution of pressure, wind, and temperature during cold, normal, and warm seasons in the United States Fall (After Fassig)

A forerunner of these later efforts and their realization is represented by the excellent study of Oliver L. Fassig, *The Climate of Baltimore*.[5] The first part of Fassig's work deals with "static climatology," the second with "dynamical climatology." There the "types of weather characteristic of the geographical horizon of Baltimore" play an important role. A large series of typical weather maps on the occasion of special storms, rain areas, frost areas, cold waves, blizzards, are published in this highly interesting book, which may serve as a pattern for future climatographies.

Figures 42 to 45 represent an interesting attempt. The average distribution of pressure, wind, and temperature of certain months (December, March, June, October) which are classified as cold, normal, or warm, is shown in the maps The four months are chosen as representative of the four seasons. Although a discussion of the maps would be out of place here, it is clear from these charts that this is the way to understand the *development* of the climatic features of a region

From these attempts, it is only one step to the numerical and graphical representation of average weather types: e.g , 20 cases of *Strong foehn* wind, east of the Rocky Mountains, are chosen. "Strong foehn" has to be defined, for example, by the condition that the temperature rises 10°F or more within one hour at a certain place, representative of the phenomenon. High floods, blizzards, hot and cold spells, dry and wet spells can be satisfactorily discussed by means of average numerical values and their cartographical representation. In the writer's opinion, such dynamical methods would arrive at the best physical synthesis of the climatic elements. In this way, a description of important and characteristic climatic phenomena could be given for great regions of the continents and for some parts of the oceans, as a guide for long range forecasting [6] and for many other purposes.

[5] Oliver L. Fassig, *The Climate of Baltimore* (Maryland Weather Service, Special Pub , vol. 2, Baltimore, 1904). See also Gardener Reed, Jr , in QJRMS 1910, p. 39.

[6] The ideas of a dynamic climatology have also been well expressed in some studies by C. F. Brooks and his pupils in "Papers in the Relation of the Atmosphere to Human Comfort" (MWR, vol. 53, 1925, pp. 423–437), containing the following articles: C. F. Brooks, "The Cooling of Man under Various Weather Conditions"; E. C. Donnelly, "Human Comfort as a Basis for Classifying Weather", F. Howe, "The Summer and Winter Weather of Selected Cities in North America"; E. S Nichols, "A Classification of Weather Types"; I. E Switzer, "Weather Types in the Climates of Mexico, the Canal Zone, and Cuba " See also E. S. Nichols, "Frequency of Weather Types at San José, Calif.," MWR, vol. 55 (1927), pp. 403, ff The investigations mentioned attempt to avoid the very complicated system of E. E. Fedoroff (MWR vol 55, 1927, pp 401 ff.) and consider the fact that the principles of the representation should not be more complex than the phenomenon itself

XVIII. 8. The Bioclimatological Part

A bioclimatological section—although a short one—is expected in a climatography. Radiation, cooling power, drying power should be mentioned. If no direct observations exist, formulae (mentioned above) enable the climatographer to give estimations of cooling–and drying–power by means of average values of temperature, wind, and humidity.

If direct observations of radiation are lacking, at least records of duration of bright sunshine and descriptions of radiation-climate or personal experiences permit a first estimation of radiation conditions.

Some words about acclimatization, endemic diseases, health resorts, etc., are usually desired in a more or less complete climatography. The immense hygienic literature presents reliable sources from which the climatographer can obtain good information without assuming too much responsibility. Every climatologist should be cautioned against giving his own opinions about subjects which are outside the range of his special realm.[7] On the other hand, personal experiences with reference to human comfort and striking phenomena in the organic and inorganic realm are always of the greatest value and should not be suppressed.

XVIII. 9. Description of Climatic Phenomena

It should never be forgotten that geography, and even more climatology, started with descriptions by early travelers. From the books of travels of the famous Venetian Marco Polo to those of Alexander von Humboldt, a nearly innumerable series of keen travelers with open eyes gave us our first knowledge of the climate of remote countries. Nor should modern climatography renounce the lively description of climatic impressions drawn from personal experience.

Puka, Albania (42.0°N, 19.9°E, 2800 ft) is located on the steep slopes of the Albanian Alps, which run along the Adriatic Sea coast from about NW to SE. Here, in summer, an average of six inches of rain falls in eight rainy days. The rain intensity is, on the average, about 0.8 in. (For comparison, in Boston the intensity is

---

[7] An instructive example of this restriction to an objective report (as far as non-geophysical influences and effects are concerned) is offered by B. Gutenberg, "Geophysik and Lebewesen," in *Lehrbuch d. Geophysik*, edited by B. Gutenberg (Berlin, 1929).

FIG. 46. Eroded mountain slopes with westerly exposure above Puka, Albania.
(42°N, 19 9°E, 2800 ft)

0.3 in.)   This numerical description is interesting and characteristic of the mountain climate along the Adriatic coast.   A glance
at Figure 46 gives life to the numbers, and we see the effect of
these downpours separated from one another by long droughts.
The slopes with westerly exposure are deeply eroded; desertlike,
barren ground is before our eyes, and that in a climate where olive,
fig-tree, and pomegranate thrive and flourish in well protected sites.

Thus the climatographer should not omit descriptions and suitable pictures which offer the means of making his statistical and
dynamical features impressive.   Books of travels, guidebooks, etc.,
often provide good contributions in this respect, if personal experience is lacking.   Tree growth and vegetation are in closest
relation to climate, and architecture also.

In the Saas Fee valley in the south of Switzerland (one of the
most beautiful valleys of the Alps), the hay barns stand on high
stone pillars.   Even if observations of precipitation did not exist,
from the height of the pillars we should know not only that copious
snowfalls occur in this region but also the average maximum depth
of snow on the ground.

In the whole Mediterranean region from Italy to North Africa,
narrow lanes are usual—so narrow that two people can hardly pass
one another.   High altitudes of the sun, combined with frequently
cloudless skies and a dry, clean atmosphere with low turbidity
factors characterize the weather of the summer half-year.   The
architectural effect resulting from these scientifically described
radiation conditions is narrow lanes.   Conversely, we conclude
that high radiation is present when narrow lanes are the characteristic architecture.   Arcades along the streets signify frequent
downpours as well as intense radiation.   High gable roofs are often
the architectural expression of habitually intense snowfalls.

High walls encircling precious orchards and vineyards are
characteristic of the northern coast of the Mediterranean, protect-
ing the crops against the mistral and the bora and similar bad and
frequent storms.   In Normandy and Brittany the traveler is
surprised to see miles and miles of high hedges, which surround
fields and farms.   One concludes, here again, that frequent storm-
like winds are found where such walls and hedges are typical.[8]

Numerous settlements high on the slopes and a deserted bottom
of the valley indicate frequent floods, and also intense inversions

[8] The reader will find excellent examples of climatically characteristic photos in the
Atlas of American Agriculture, in the sections on "Soil" and "Natural Vegetation."

of temperature in fall and spring.   Manners and customs of man-kind are often highly influenced by climatic features.   It was natural that the traveled R. DeCourcy Ward was enthusiastic about the necessity of colorful descriptions; he went so far as to speak of a "car-window climatology."   None other than J. Hann always enjoyed getting lively descriptions to cover his numerical skeletons.   The many descriptions in his famous regional climatography of the world attest these efforts.

CONCLUSION

THE READER who runs over the pages of this book is perhaps disappointed in not finding this or that special method.

In the first pages, it was shown that the number of derived elements is not limited theoretically; thus completeness is unattainable.[1]

Moreover, it is intended to present only a system of methods which facilitates the step from the qualitative to the quantitative. Many ways are offered for attaining this purpose, and they all should lead not only to an exact and comparable description of the climate but, in the end, to a physical explanation.

[1] The reader will find a rich source of different methods of representation in the monumental work· *Atlas of American Agriculture Physical Basis including Land Relief, Climate, Soils, and Natural Vegetation of the United States* (Washington, 1936), prepared under the supervision of O E Baker, with contributions from the Weather Bureau, Willis R Gregg, H. G. Knight, Frederick D. Richey, F A. Silcox, A G Black.

APPENDIX I

FORMS OF INDIVIDUAL CLIMATIC TABLES

THE FOLLOWING form is that of the "Climatic Summary of the United States" published by the U. S. Weather Bureau. The form is slightly modified with respect to its heading. Two columns have been added: "Depth of Snow on the Ground" and "Estimated Average Wind Velocity." The first columns can be filled with information, based on the regular observations of cooperative stations in the United States

The average wind velocity is not recorded by the cooperative observers. An estimate of the average of the whole day could easily be made according to the following simplified scale:

$$0 = \text{calm to light}$$
$$1 = \text{gentle}$$
$$2 = \text{strong to stormy}$$

Too little attention is paid to the wind in climatological publications.

BLANK FORM OF A CLIMATIC TABLE

(Taken from the "Climatic Summary of the United States," U S Weather Bureau)

Name of the Place . . . . . Country . .

$\phi =$ . . $\lambda =$. $\textit{B} =$ .. feet (meters)

Period . . . .

[Remarks about special local conditions, period, instruments, etc]

* According to daily estimations: 0 = calm to light, 1 = gentle, 2 = strong to stormy.

216

Julius Hann suggested the following *content* of a climatological table The form is in principle identical with that of the U. S. WeatherBureau. A great number of climatologists, perhaps the majority, have made use of the following suggestions as far as available data permitted

A. Temperature
   a. Averages for the three observation times
   b. True monthly means
   c Extreme monthly and annual means for the period in question
   d Averages of the interdiurnal variability
   e Mean daily periodical range (if records of a thermograph are available)
   f Mean daily extremes
   g. Mean absolute extremes and their difference
   h. Absolute extremes
   i. Average frequencies of certain differences between consecutive daily means (e.g differences 3°F to 6°, 7° to 10°F etc.)
B. Vapor pressure
C. Relative humidity (morning, early afternoon, daily mean)
D Cloudiness (morning, early afternoon, daily mean)
E. Sunshine duration (hours, percentage)
F. Precipitation
   a. Average amounts
   b. Relative variability
   c. Snow (and snow mixed with rain), amounts
   d Depth of snow on ground
G Wind
   a Resultant direction
   b Resultant velocity (vectorial)
   c Average velocity, disregarding the direction
   d Steadiness
H. Air pressure
   a. Averages
   b. Average monthly and annual range
I. Number of days
   a. Clear   Overcast
   b Without sunshine
   c With precipitation ≧   ·
   d. With hail
   e With snowfall
   f. With thunderstorm
   g. With storm (> Beaufort 6)
   h. With frost (minimum below freezing point)
   i. With fog

This table should be completed at least by a frequency distribution (per cent) of 8 wind directions and calm for the months and the year. Other frequency distributions could be added according to the purpose of the climatography.

APPENDIX II

$$\sqrt{\frac{1}{n}}$$

(See XII 2 b and the "Intermediate Chapter")

	0	1	2	3	4	5	6	7	8	9
0		1.000	.707	.577	.500	.447	.408	.378	.354	.333
10	316	302	289	.277	267	258	250	243	236	229
20	.224	218	213	209	204	.200	.196	192	.189	186
30	183	.180	177	.174	171	169	.167	164	.162	160
40	.158	156	154	.152	151	.149	.147	146	144	.143
50	141	.140	139	.137	136	135	134	132	131	130
60	.129	.128	127	126	.125	124	123	122	.121	.120
70	120	119	118	.117	116	115	115	114	113	.113
80	112	111	110	.110	109	108	108	107	.107	.106
90	.105	.105	104	.104	103	.103	102	.102	101	101
100	100	.100	099	.099	098	.098	097	097	096	096
110	095	095	.094	.094	094	.093	093	092	092	092

APPENDIX III

AUXILIARY TABLE, CALCULATING THE EQUIVALENT TEMPERATURE FROM AIR-TEMPERATURE ($t°C$) AND RELATIVE HUMIDITY (AFTER F LINKE)
(See IX.2)

$t°C$	$K(t, b)$ Air-Pressure in Millimeters of Mercury				
	780	760	740	720	700
−45	0 11	0 11	0.12	0.12	0 12
−40	0 20	0 20	0.21	0.22	0.23
−35	0 33	0 34	0.34	0.35	0.36
−30	0 57	0 59	0 61	0 62	0 64
−25	0 95	0 97	1 00	1.02	1.05
−20	1 55	1.59	1 64	1 68	1.72
−18	1 87	1 92	1 98	2 04	2.09
−16	2.29	2.34	2.40	2 47	2.52
−14	2 75	2 81	2 88	2 97	3 02
−12	3 26	3.37	3 45	3 54	3.61
−10.	3 90	4.01	4 11	4 23	4 33
− 8	4 72	4 87	5 00	5 12	5.26
− 6	5 80	5 87	6 00	6 20	6 40
− 4	6 77	6 90	7.13	7 33	7 60
− 2	7 90	8 00	8 25	8 50	8 75
0	9 09	9 32	9 58	9 84	10 1
2	10 5	10 7	10.9	11 2	11.5
4	12 0	12 2	12 5	12.9	13.3
6	13 9	14 2	14 6	15 1	15 5
8	15.8	16 2	16 6	17 2	17 6
10	18 1	18 6	19 1	19 6	20 1
12	20 6	21 1	21 7	22 3	22 9
14	23 4	24 2	24 8	25 5	26.2
16	26 6	27 3	28 2	28 9	29.7
18	30.2	30 8	31.7	32.7	33 6
20	34 2	35 1	36.1	37 0	38 2
21	36 4	37 2	38 4	39 4	40.4
22..	38 6	39 5	40.6	41.7	43 0
23	40 8	41 9	43.4	44 5	45 6
24	43.5	44.5	46 0	47 3	49.4
25	46 7	47 5	48 8	50 2	51.6
26	49 0	50.1	51.6	53.0	54 6
27	51.9	53.2	54.6	56 9	57.8
28	54 9	56 3	57.9	59 6	61.2
29	58 3	59 7	61 4	63 0	64.7
30	61.9	63 5	65 2	67.0	68 8
31	65.4	67 2	69 1	71 1	72 6
32	69 2	71.2	72 9	75.2	76 7
33	73 3	75 2	77.1	79.3	81 3
34	77 6	79.6	80 4	84.0	85 0
35	82	84	86	89	91
36	87	89	91	94	96
37	92	94	96	99	102
38	97	100	102	105	108
39	103	105	108	111	114
40	108	110	113	116	120

APPENDIX IV

The Days of an Ordinary Year, Numbered Consecutively, Beginning January First

	Jan	Feb	Mar	Apr.	May	June	July	Aug	Sept	Oct.	Nov	Dec
1	1	32	60	91	121	152	182	213	244	274	305	335
2	2	33	61	92	122	153	183	214	245	275	306	336
3	3	34	62	93	123	154	184	215	246	276	307	337
4	4	35	63	94	124	155	185	216	247	277	308	338
5	5	36	64	95	125	156	186	217	248	278	309	339
6	6	37	65	96	126	157	187	218	249	279	310	340
7	7	38	66	97	127	158	188	219	250	280	311	341
8	8	39	67	98	128	159	189	220	251	281	312	342
9	9	40	68	99	129	160	190	221	252	282	313	343
10	10	41	69	100	130	161	191	222	253	283	314	344
11	11	42	70	101	131	162	192	223	254	284	315	345
12	12	43	71	102	132	163	193	224	255	285	316	346
13	13	44	72	103	133	164	194	225	256	286	317	347
14	14	45	73	104	134	165	195	226	257	287	318	348
15	15	46	74	105	135	166	196	227	258	288	319	349
16	16	47	75	106	136	167	197	228	259	289	320	350
17	17	48	76	107	137	168	198	229	260	290	321	351
18	18	49	77	108	138	169	199	230	261	291	322	352
19	19	50	78	109	139	170	200	231	262	292	323	353
20	20	51	79	110	140	171	201	232	263	293	324	354
21	21	52	80	111	141	172	202	233	264	294	325	355
22	22	53	81	112	142	173	203	234	265	295	326	356
23	23	54	82	113	143	174	204	235	266	296	327	357
24	24	55	83	114	144	175	205	236	267	297	328	358
25	25	56	84	115	145	176	206	237	268	298	329	359
26	26	57	85	116	146	177	207	238	269	299	330	360
27	27	58	86	117	147	178	208	239	270	300	331	361
28	28	59	87	118	148	179	209	240	271	301	332	362
29	29		88	119	149	180	210	241	272	302	333	363
30	30		89	120	150	181	211	242	273	303	334	364
31	31		90		151		212	243		304		365

INDEX